SEA LIFE IN FOCUS

Sea Horses (Hippocampus guttulatus)

SEA LIFE IN FOCUS

A Memoir

DOUGLAS P. WILSON

DSc, CBiol, FIBiol, HonFRPS

with a Foreword by
ERIC HOSKING
OBE, FIIP, Hon FRPS

cell mead press

Published in 2002 by
Cell Mead Press, 60 Church Road, Old Windsor, Berkshire, SL4 2PG

Copyright © DPWilson Ltd, 2002

All rights reserved. No part of this publication may be reproduced, stored in a retrieval system, or transmitted, in any form, or by any means, electronic, mechanical, photocopying, recording or otherwise, without the prior permission of the publisher or DPWilson Ltd.

ISBN 0 9526678 1 9

Front cover photograph: Rosy Anenome (*Sagartia elegans*)

Back Cover (*top*): Lesser Octopus (*Eledone cirrhosa*)
(*bottom*): Larva of Long-armed Starfish (*Luidia sarsi*) metamorphosing

Printed by Creeds the Printers, Broadoak, Bridport, Dorset, DT6 5NL

To my Grandchildren:

Imogen, Damian, Marcus, Olivia, Melanie
and Alison

and in memory of her namesake
Alison, my wife

Editor's Note

In the summer of 2001 the typewritten manuscript of my father's memoirs came to light again, more than a decade after it was written. As I re-read it I thought how engagingly it had been written, communicating enthusiasm for the subject and packed with interesting observation and anecdote. It surely deserved better than to languish at the back of a cupboard. Years earlier, in 1987, the late distinguished bird photographer Eric Hosking tried to interest a publisher in it but without success. The trouble apparently was that it wasn't about birds: the general public was not thought to be interested in sea life.

That may have been true in the 1980s but is surely not so now, especially after the BBC television series *The Blue Planet*. My father would have been fascinated by it and full of admiration for the courage and skill of the photographers. He would also have reflected that the remarkable images they brought back, whether from coral reefs, polar seas or the dark depths of the ocean, are a very long way from his own efforts in the 1920s and 30s to achieve simple black and white shots of cod or mackerel in an aquarium. Yet marine photography had to begin somewhere, and his perseverance and inventiveness in achieving the then seemingly impossible had its own element of heroic enterprise.

Sea Life in Focus surely makes a contribution to photographic history, and also shows that much more went into the photographs than technical ingenuity. These are not just snapshots but portraits of the living animals. Their artistic excellence was recognised in the many awards my father received in his lifetime, including the Honorary Fellowship of the Royal Photographic Society; it is so still in their continued publication in books and journals around the world.

In editing these memoirs I have been helped and encouraged by my brother Richard and other family members. I am also extremely grateful to David Hosking for his generous help with the colour photographs. As a non-biologist I have tried to be accurate in transcribing scientific names: for any errors I crave indulgence.

Hester Davenport

Contents

Foreword by Eric Hosking ix

Introduction 1

1. Penny Cameras 3

2. Photographic Challenge 13

3. Fishes by Flashlight 20

4. Microscopic Monsters 34

5. Underwater Stories 47

6. On the Seashore 64

7. Strange Strandings 79

Afterword 91

Bass (Dicentrarchus labrax)

Foreword

When I urged Dr Douglas Wilson to write this book I had no idea that he would retaliate by asking me to write the Foreword! But I have read it with great interest. Douglas is not only a very skilled photographer but also a highly respected scientist, and it is a delightful account of photographic techniques and amusing personal reminiscences, mixed with some extraordinary and fascinating accounts of sea life. I was not surprised to learn that some of his photographic results revealed new discoveries about the marine world.

When Douglas and I met before the war, photography was in many ways much more interesting and exciting because you did it all yourself. My wife often jokes that we never had a honeymoon though we spent three months in the Scottish highlands. After a day's photography I would never dream of going to bed until I had developed all the negatives I had taken that day – by the time it was done it was well after midnight and my bride was sound asleep!

In those days if you required a piece of apparatus to do a particular job it was more than likely that you had to build it yourself. Douglas describes vividly some of the problems he had and, as he says, some of the things he made might have been created by Heath Robinson. Days would be spent experimenting with bits of Meccano, parts of old electric bells, pieces of metal and wood, but in the end a gadget was produced that worked – and the photographer could feel a justifiable pride in his achievement.

It was surely owing to Douglas's ingenious devices that he was able to obtain successful marine photographs where others at that time had failed. He was almost certainly the first in the world to photograph living plankton under the microscope by adapting a specially made high-speed flash unit. But even then what a feat it was taking a picture! He had to use both hands to manipulate the tiny creatures into the right position and keep them in focus, and expose the film by pressing a connection to the camera shutter with his foot.

Sea Life in Focus

I remember that when the *New Naturalist* series of books was conceived in 1943, I was invited to become photographic editor. It was my job to enlist some of the outstanding wildlife photographers of the day to take colour photographs especially for it. One of the first I contacted was Douglas who was really enthusiastic, partly, I think, because we would supply the colour material, 9 x 12cm Kodachrome (long since discontinued unfortunately), which was unavailable to the public since it was manufactured in America, where it had to go back for processing. His results were outstanding and almost all the photographs that appeared in Professor C.M.Yonge's *The Sea Shore* were taken by him. Many other of his photographs appeared in Sir Alister Hardy's two volumes *The Open Sea*, and in years since countless more have been reproduced in books and magazines all round the world. Here in this book his keen eye for a photograph can again be seen, and the skill with which he captured the darting, scuttling creatures admired once more.

Eric Hosking

Photo by Eric Hosking of DPW as he prepares to take the picture of sea horses which forms the frontispiece

Introduction

It was my friend Eric Hosking the eminent bird photographer, who suggested the writing of this book, on rather similar lines to his own *Eric Hosking's Birds* (1977). After initial doubts I thought it over and looking again at his book which recounts his early photographic efforts with relatively primitive apparatus felt that I might be able to write something of the sort. Not about birds, of course, but about marine life with which I have been engaged both professionally as a research worker in one of the world's leading marine biological laboratories, and as an amateur photographer with a passion for photographing creatures of our seas and shores.

When I first took an interest in marine life, books on the subject were largely illustrated with black and white or coloured drawings. When photographs were reproduced they were much more likely to be of dead, often preserved, specimens than of living animals. There were a few notable exceptions: some books, such as Martin Duncan's *The Sea Shore: A Book for Boys and Girls* (1912), had photographs of both dead and living seashore animals. One little picture book, *Seashore Life* by W.B. and S.C. Johnson (1907) had sixty photographs of living animals and plants taken on the seashore or in a tank. But the outstanding books were those by Dr Francis Ward, *Marvels of Fish Life* (1911) and *Animal Life under Water* (1919). Ward photographed both freshwater and marine animals and his work was marked by tremendous efforts to ensure that he portrayed his fishes and other aquatic animals in normal environments and in natural light. To achieve his aims with freshwater life special ponds were constructed with side observation chambers below water level. Vertical windows looked out on the underwater scenes, and some had angled windows to avoid distortions. The chambers had roofs to pull down once the photographer was inside, ensuring that no light entered except from the pond, thereby avoiding reflections and concealing the observer from the fishes, which saw only their own reflections in the glass. The resulting photographs, some on Lumière colour plates, of fishes, otters and diving birds are truly remarkable, as are also pictures showing how a fish would see an angler fishing from the bank.

I remember my father taking me to a lecture by Dr Ward in Manchester soon after the First World War. On the platform he arranged a row of chairs to represent a hedge and with a fishing rod demonstrated how an angler could avoid being seen by the fish. Crawl low in front of the hedge, as he crawled low in front of the chairs, keep the rod low too and the fish won't see you – stand up and cast your line from over the hedge and it will! His photographs, taken through one of his pond windows, show just what a fish would or would not see from those positions. Other photographs reveal how in their natural environment the silvery undersides of fishes reflect the colours of their surroundings, making them almost invisible in side view.

For some freshwater and marine work Ward used specially constructed tanks, again taking his photographs with sky lighting only. He seems not to have used any form of artificial lighting, except for photomicrography. He also waded in shallow coastal water, his reflex camera within large square-section 'tubes', with lower ends submerged to avoid reflections and give clear views of the bottom. His observational and photographic work was remarkable for its time – all being done before 1919 – work which has never received its full acknowledgement. Though I too have always aimed to depict marine life as naturally as possible my own approaches to the problems have been different; I could never have afforded to adopt his methods even had I been fully aware of them. As I hope this book will show, my methods had to be commensurate with what I could afford.

I am much indebted to my daughter, Hester, for her encouragement, and for many hours of typing and editing the text. Richard, my son, also made some useful suggestions. To Eric Hosking and his son David I am indebted for much help. My only sadness has been that my wife Alison is no longer alive to help me: her wide knowledge and experience would have contributed much to the book. Nevertheless, what I recount owes much to the help and encouragement she gave me throughout the years.

CHAPTER ONE

Penny Cameras

B ut with six cameras I could take six photographs.' The year was 1913, the place a toy-shop in Alexandra Road, Manchester, and I had sixpence to spend. It was just after my eleventh birthday, the sixpence a present – quite a lot of money in those days. Earlier I had bought from the same shop a penny pin-hole camera, a little black cardboard box, one half sliding within the other, the smaller half perforated at one end with a minute hole surrounded by a cardboard collar on which was a tiny cardboard cap. Inside, wrapped in black paper, was a small glass plate coated with sensitive emulsion, packets of developing and fixing 'salts' and an instruction leaflet – all for one old-fashioned penny. With this outfit I had photographed the back of our house, with the camera jammed in a fork of a large pear tree to enable an exposure of many seconds to be made. The plate was developed and fixed in saucers on the stone steps leading down to our cellars. With the doors closed all was completely dark. What a thrill it was when the plate was brought into the light to see not only the house outlined against the negative darks of the sky, but the windows and even the window bars!

So it was that when I had that sixpence to spend I hurried back to the shop for more cameras, resisting the pressure of the owner to sell me a larger and better camera, price sixpence. I pointed out that with it there were only two plates, while the smaller penny cameras had one each. What I did with that clutch of cameras I have only the vaguest recollections; I do know, however, that the negatives were never printed. A cardboard printing frame with sensitive daylight paper cost one halfpenny and I had spent all my money.

My interest in photography had been aroused at an early age. I remember being taken, when about three years old, to a local photographer

Douglas and Vera watch the birdie

to pose for portraits dressed in a sailor suit. The studio was lit only through an overhead glass roof and the camera was a huge contraption on a massive stand. The photographer, just before taking off a cap from the lens, said 'Keep still, watch for the dicky-bird.' 'Didn't you see it?' he asked, 'you must have missed it.' This episode left me puzzled and intrigued. I remember, too, having my head steadied from behind with a head-clamp, often used in those days of long exposures; also being posed with my younger sister sitting on a great stuffed St Bernard dog. Those photographs are still preserved in the family album.

A few years later I was photographed with a fish: prophetic of things to come! The photographer this time was a keen amateur, who lived opposite us in Mayfield Road. I was taken across one day with my sister to be photographed in his garden and I insisted on my fish going with me – a lovely golden celluloid fish, a great treasure which I held on the upturned palm of my hand. Afterwards I was allowed into the cellar dark room to be shown how the plate was developed. It was a mysterious place, lit by a dim red lantern standing on a bench which ran along one wall, with a bats-wing gas burner above it. The photographer explained the reason for the red light and the darkness as he turned down the burner until it was nearly out. I watched as the image, scarcely visible in that dim light, slowly appeared on the plate rocked in a dish, and it was exciting to see a tiny dark fish-shape appear on the image of my hand. A day or two later we were given some prints; there was the fish, but the print was rather foggy, as was all that photographer's work. Years later I realised what had happened: he should have turned out the bats-wing burner completely, for even very low it was giving sufficient light to fog the plate in the developing dish below. But I wish I still had that print!

After my first excursions into the technique of photography it was towards the end of World War One before I again acquired a camera, this time one with a lens, a Box Brownie. The size was chosen with care after I had several times flattened my nose against the window of a chemist's shop in Rhos-on-Sea in North Wales during one of numerous holidays with an aunt who lived there. In the window of the shop was a display of Box Brownies and other cameras of various sizes with alongside each a sample photograph. I chose a landscape format – 4¼ x 2¼ ins – and later, as soon as I had enough money (fourteen shillings and sixpence if I remember rightly) I bought the camera one day back home in Manchester, which was rather hard on the North Wales chemist. The next morning I dragged my sister's terrier, Jack, to a patch of sunlight on the lawn, made him lie down, and took his portrait. This was my very first photograph using a camera with a lens.

During the next year or two the Brownie produced a lot of pleasing photographs printed in the garden on daylight self-toning paper. What fun it was in those days to spread the wooden printing frames along the garden path and watch the positive image gradually darkening until done enough. The printing paper was then washed for toning in a dish in the kitchen sink, followed by fixing in a bath of hypo. I still have many of those lovely sepia prints. Young photographers miss a lot now that daylight print-out papers with their beautiful brown tones are unobtainable.

However, the Brownie had its limitations; I very much wanted to take close-ups of the exciting creatures I was meeting on the seashore in Penrhyn Bay during long summer holidays staying with my aunt at Rhos-on-Sea. While still at school I was scouring the shops for such books about sea life as I could afford. No other subject fascinated me so much; I wanted above all to be out on the shore and I wanted to

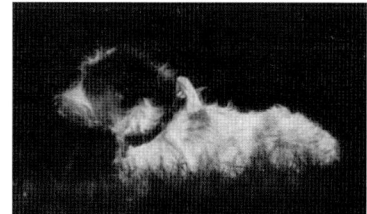

Jack

Sea Life in Focus

Penrhyn Bay in North Wales, looking towards the Little Orme

photograph what I saw there. So I sold the Brownie and with what it fetched and some savings bought a second-hand wooden half-plate camera with double extension, Rothwell R.R. lens f/8-f/64, and three wooden double dark slides – all for thirty shillings. A wooden tripod costing six shillings is still a treasured possession, in use for over sixty years, the sliding legs just as they were. I would rather lose an expensive replaceable single-lens reflex than those legs, so much more useful in rock pools than corrodible metal legs. Wooden legs of that type seem no longer to be made. Later a roller-blind shutter which fitted over the front of the lens was added to the equipment. So with camera and double dark slides wrapped up in a black focussing cloth in a knap-sack on my back, the lens in a padded bag in one pocket, the shutter in a box in the other, an exposure meter and the tripod with essential tripod screw, I could explore the shore and begin to record the life of the sea.

Penny Cameras

Penrhyn Bay is a relatively easy shore to wander over, though not for the bare-footed. Wide fairly level areas of round stones and pebbles bound together by the byssal threads of mussels growing on and between them stabilise the shore, preventing the surface from washing away. Here and there, concentrated in some areas, are huge glacial boulders from the last Ice Age. They are covered with acorn barnacles, dog whelks, periwinkles and so on; some are pitted with small pools in which corallines and various animals live. These boulders are firmly bedded in the ground, immovable in the stormiest weather, and the last time I saw them, in 1978, they were in the same positions as when I was a boy. Depressions in the stony areas form shallow pools one can walk around, and there are large sandy areas close to the low tide mark. At the lowest levels, uncovered only by spring tides, are more boulders covered with seaweeds and sponges not to be found at higher levels. Such is the shore I used to wander over with my camera and tripod, often in ankle-high laced-up leather boots which hardly ever got wet, later in sandals which did.

I couldn't afford half-plates. Out of plywood I made adaptors to take ¼-plates, loaded and unloaded in a changing bag on my aunt's table. The plates were orthochromatic, Imperial Special Rapid mostly, pretty slow by present day standards. Exposures were calculated with a Watkins Bee Meter, in size and shape like a small pocket watch. A small portion of a disc of sensitive paper was exposed and seconds counted until it had darkened to match a standard tint alongside. In sunshine this took about two seconds, in a dim light it could be minutes. From the reading a rotating scale gave exposure time according to emulsion speed and stop used. As I had no watch much depended on accurate counting, and the allowance made for camera extension with close-ups. In poor light it was best to select a stop with which the exposure time would equal the meter darkening time. The meter would then be placed alongside the subject just out of camera view and the sensitive paper uncovered as soon as the cap was taken off the lens. Of course, this method only worked with static subjects; luckily many shore animals such as barnacles, limpets, periwinkles and mussels are static at low tide. Even starfishes and crabs can often be relied on to keep still for a few seconds. In those days most exposures were made by taking the cap off the lens and putting it on again a second or two later. Only occasionally was the roller blind shutter used. Whenever possible the lens was stopped down to f/64 to obtain as great a depth of focus as possible. With that antique lens such small stops did not noticeably degrade the image.

With at most six exposures possible on any one visit to the shore there had to be a careful selection of subject matter. I would wander over wide boulder-strewn and sandy areas looking for something worthwhile to photograph, an unusual animal or a pleasing group of limpets, periwinkles or other shore organisms. I quickly learned not to put together artificial groupings, or to rearrange seaweeds; what was photographed had to be as the sea left it. At most it was permissible to remove bits of washed-up debris, but never to adjust the position of a periwinkle or other organism, thereby falsifiying the ecology. This principle has been adhered to ever since when photographing on the seashore. However, it has always been necessary with active animals, such as crabs, which scuttle away when disturbed from under a stone, to get them to settle down in some convenient place, and creatures which burrow in sand or mud can rarely be photographed in their living quarters.

Since I had no watch I got into trouble many a time for being late for meals: aunts don't like to be kept waiting. On a morning low tide I could judge the time to leave the shore by the appearance on the horizon of the excursion steamer 'La Marguerite' on

The intrepid young photographer

her way from Liverpool to Llandudno and the Menai Straits. When she came into view I knew it was about half-past twelve, time to leave for one o'clock lunch. One August day she did not appear, delayed, I believe, by mechanical trouble. Eventually hunger forced me to realise that it must be later than I imagined and I hurried away to find an anxious aunt on the look-out from her side door, convinced that something dreadful had happened – I had fallen down and broken a leg, or drowned in a rock pool. Her relief at seeing me safe and well mitigated to some extent what she said to me over lunch.

When holidays were over the exposed plates were taken home to Manchester for developing and printing in my gas-lit attic dark room. The room gas-bracket was inconveniently sited for making contact prints on gas-light paper, or for making lantern slides, and so I fixed a secondary bracket to a wall some distance away, bringing gas to it from the room bracket with rubber tubes and connectors. The arrangement would give a modern fire-prevention officer a fit! The safe-light had its own paraffin lamp. It was tiresome having to carry all water from and back to the bathroom on the floor below. Once I incurred parental wrath when I had the bright idea of emptying the wash water out of the window onto a garden bed below.

A few of the early photographs later proved to be of some biological interest. A picture of Common Periwinkles (*Littorina littorea*), for example, illustrated one of their habits. While drying out in the sun they produce a little mucus which sticks the lips of their shells to the rock. The mucus hardens, enabling them to retreat within their shells to keep moist. On vertical or steeply sloping surfaces the hardened mucus holds them in place only when the shell lip is uppermost. And so row upon row of crowded periwinkles are often seen all suspended in this same way. Just occasionally one mistakenly fixes itself with the shell lip downwards and topples over. The first photographs I ever had reproduced were of periwinkles illustrating an article in the *Journal of the Manchester University Science Federation* (1925), and one showing the suspension just described appeared in the science magazine *Nature* in 1929.

Manchester Microscopical Society, Whitsun 1921, DPW at the door

My photographic explorations on the shore had scarcely begun when on leaving school in 1919 I was employed, first as office boy, later as junior shipping clerk, in a small Manchester office shipping cotton piece goods to India. My former long holidays in Rhos-on-Sea were now curtailed to about a week at Whitsun and ten days in the summer. But one memorable Whit Week, in 1921, I accompanied several members of the Manchester Microscopical Society to the Marine Station at Piel near Barrow-in-Furness. It was there under the tuition of the resident superintendent, Andrew Scott, that among other marvels I saw living marine plankton for the first time, and learnt how it could be collected in a conical net of fine mesh let down into a current flowing under a jetty. As soon as I got home I made, with mother's help, a similar tow-net of muslin, its wide mouth sewn round a hoop. I knew that at Rhos-on-Sea the falling tidal stream ran strongly under the pier (demolished in 1954), and that by letting the net down into the current on the end of a long strong cord, it would collect plankton.

On my next visit I eagerly tried it out. It cost more to fish from the pier than to stroll along it, but the kindly pier-master, Mr Chubb, allowed me on without paying the extra as I was not intending to catch fish. It was amusing, and sometimes a little embarrassing as the net was hauled in after twenty minutes or so, when people on the pier rushed to see the large fish they thought had been caught. As the net was turned inside out to wash off into a jar of sea water the brownish slime of plankton concentrated in its tip, there were sometimes exclamations of disappointment, sometimes demands for an explanation, and occasionally an attentive audience for an impromptu lecture on the composition of plankton and its economic importance. Leaving the pier I would show my jar of minute darting specks to Mr Chubb and on one or two occasions took a borrowed microscope along with me to interest him in his office at the head of the pier with a closer look at some of the living treasures in the jar. Back in my aunt's house a happy hour or two were spent searching through the catch while still alive with the microscope, then it was pickled in formalin. Eventually there was a considerable accumulation of bottles of preserved Rhos-on-Sea plankton, which later proved of some value to Dr Marie V. Lebour of the Marine Biological Laboratory, Plymouth, when writing her Ray Society Monograph 'The Marine Plankton Diatoms of Northern Seas'.

At home in Manchester, away from the sea, there was little opportunity for practical marine work. I attended meetings of the Microscopical Society and as the youngest member received much help and many kindnesses from its members, one of whom had lent me the microscope. One of their over-riding interests was pond life and that, being aquatic, interested me too. But of greater significance to my future career were the evening extra-mural lectures in Zoology given by dear old Dr Stuart Thompson in the Zoology Department at Manchester University. Attended largely by members of the Microscopical Society there was once a week through the winter a fascinating lecture on the anatomy of some type of invertebrate, followed by an hour's practical dissection in the laboratory. For the first time in my life I knew what I really wanted to do – but I was stuck in that shipping office, with little prospect of ever doing anything else. Then events beyond the control of anyone in this country intervened.

Sea Life in Focus

In India the Mahatma Gandhi, with his campaign for 'home spinning and weaving', was slowly but surely ruining those parts of the Lancashire cotton industry engaged in trade with India. Manchester shippers were facing hard times, staffs had to be reduced and often it was 'last in, first out'. I was a 'last in', living at home without family responsibilities, and so, in spite of the best my boss could do to get me another job, I was 'out'.

The world seemed to have come to an end – but not for long. My aunt generously offered to pay university fees if my parents, who were not well off, could keep me at home: there were few university scholarships or grants in those days, and none I could get. Unfortunately I had left school without university entrance qualifications, and so for a year I studied hard to pass Matriculation, which the summer after my 'sacking' I did. Shortly after my twenty-first birthday I was accepted for an Honours course in Zoology, under Professor S.J. Hickson, eventually graduating with First Class Honours and with a Graduate Research Scholarship. This latter, plus an additional grant, took me to Plymouth in September 1926. Earlier that year I had attended a students' Easter Vacation course in Marine Biology at the Laboratory, and had met the Director, Dr E.J. Allen, who kindly invited me to work under him in the Laboratory if my examination results in the summer were satisfactory. Surely the Mahatma Gandhi must be numbered among the chief of my benefactors! Without him, and at that crucial time, I should not now be writing this book.

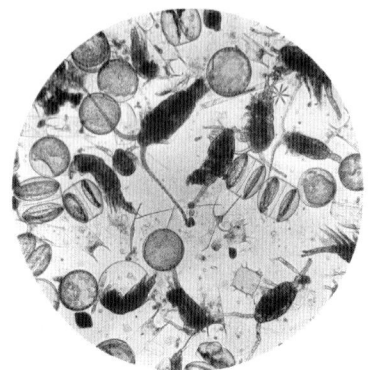

Pickled plankton

CHAPTER TWO

Photographic Challenge

My first task on arrival at the Plymouth Laboratory in the autumn of 1926 was to learn how to identify polychaetes, bristle worms, of which some 260 species were already known for the Plymouth district. They constitute an important element of the marine fauna, many as food for other organisms including commercially valuable fishes. Some, such as the tube-dwelling fan-worms, are beautiful, some are brilliantly coloured, while many others have strange habits or fascinating life-histories. Dr Allen, the Director, was himself an authority on this group and he was particularly anxious that more information concerning their breeding and larval development should be obtained.

At that time knowledge of the subject was much more imperfect than it is now, and it was to be my main task for many years to add to what was already known, especially of the developments of those species whose larvae grow while swimming feeely in the plankton, sometimes for weeks, even months. I had to discover the breeding seasons, to collect mature males and females, induce them to shed their sperms and eggs and make fertilisations in the laboratory, then to rear the larvae resulting from those fertilisations. This work I embarked on with enthusiasm, for I was already interested in the larvae from plankton-collecting days at Rhos-on-Sea, where I had so often seen interesting-looking worm larvae about which I could obtain only scanty information.

Peacock worm (Sabella pavonina)

Sea Life in Focus

The Marine Biological Association Laboratory, Citadel Hill, Plymouth

Some more interesting aspects of this scientific work will be referred to later but I want here to write primarily about photography and the various crude devices invented to enable quite ordinary cameras to take pictures of aquarium fishes and other creatures, and eventually living plankton, the latter posing a challenge not solved for many years. All this photographic work was purely a hobby; it took second place to the researches and other work I was employed to do. The great advantage was that in an environment of marine research surrounded by living marine animals of all kinds I was able to make use of the situation, especially at weekends.

Dr Allen was himself interested in photography and with a friend had tried to photograph actively-moving plankton organisms seen through a microscope. At that time it was probably an impossible task. He encouraged me in my photographic efforts, and it was a request from him that put me on the road to taking photographs of animals in aquarium tanks. He wanted me to make a photographic record of living plaice, their feeding and growth rate being then a subject of research by another member of the Laboratory staff. The fish were being reared in special cages submerged in a sheltered 'dock' over on the Cornish side of Plymouth Sound. They could be brought into the Laboratory for photography only on fixed days, whatever the weather; were it bright sunshine the job would be fairly easy out of doors, but if dull or raining, impossible. I knew that indoors we had insufficient light; as we paced up and down in a large room of which one wall was lined with innumerable bottles of preserved specimens, trying to resolve this problem of recording living ones, I suddenly said 'Flashlight!' Dr Allen came to an abrupt halt, thought a moment, then saying 'You have it,' turned on his heels – and left me to work out the details. So were born my first excursions in photography with magnesium flashlight powder.

I must go back a bit. At that time (1928) I was already an Assistant Naturalist on the staff of the Laboratory. A year on a research grant followed by a Student Probationership had taken me to the spring of that year. On 1 April I was summoned to Dr Allen's office to be informed that the Council of the Marine Biological Association had agreed to his suggestion that I be appointed to a vacancy on the research staff. 'I don't know who today is the greater fool, me or you,' he said laughingly as I gladly accepted the post. To my great joy one of my new jobs was to supervise the Aquarium to which the public were admitted – a job I was to retain until my retirement forty-one years later. This meant that I had a freer hand in the Aquarium and greater opportunity to borrow animals for transfer to photographic tanks than would otherwise have been the case. This privilege was never abused; the well-being of the animals always came first even if it meant the loss of some photographs.

It was not long after I had been given responsibility for the Aquarium that Dr Allen asked me to visit a country house to inspect a freshwater fish tank which the owners had offered to sell. It was a metal-framed tank with a slate bottom and glass windows on all four sides, and I bought it for the handsome sum of £2. This tank proved very useful not long afterwards when a film director and small camera crew came to film live fishes. They were making one of the first ever documentary films: at that time films were silent monochrome dramas and comedies accompanied by the thumping out of tunes on a piano, the player trying to capture in sound the mood of the moment as the story unfolded. The director was John Grierson and the film was *Drifters*, a revolutionary documentary first shown in 1929, recording the hard and dangerous work of those who manned the North Sea herring fleet. A drift-net fishery also flourished at Plymouth in winter months, until the herring stocks were depleted in the early 1930s, and I used to watch the fleet, which had come from Scotland and the East coast, leaving Sutton Harbour in late afternoon and setting out through the Sound to fish all night well out to sea, returning in the morning to land their catches at the fish market.

Grierson wanted to photograph swimming herring to add a little undersea scenery to his film (underwater cameras and scuba diving were unheard of then). Unfortunately we had no herring, since attempts to keep them alive in the Aquarium had always failed. However, we did have a shoal of rudd in a large freshwater tank, and it was decided that if filmed in black and white and shown only briefly, it would take a smart fishery expert to notice the deception. The normal daylighting of the Aquarium was too dim for the film being used and no floodlights were available. So it was that the new tank was cleaned and set up in the outside yard, well lit from the sky above and filled with a hose from a freshwater tap. I had the job of supervising the netting of some rudd and their transfer from the public Aquarium; after they had settled down the camera man took pictures with a ciné camera on a robust tripod. Grierson then wanted to add some authentic marine life to the pictures and selected a small conger eel to be dropped into the tank at a signal from him. It was filmed among the rudd for some seconds before being rescued and put back into sea water, none the worse for its experience. Subsequently when I saw the film I did not think those 'underwater' scenes looked too unnatural.

I played no direct role in the filming of *Drifters*, but during my probationership I was thrilled to be asked if some of my photographs of seashore life could be used in a book on which two of the Laboratory staff, F.S. Russell and C.M. Yonge, were collaborating. *The Seas* became world-famous and both men were later knighted for their contributions to marine biology. Twenty-nine of my photographs were used, several being coloured by W.J. Stokoe, so well that they almost equal present day colour ones, and indeed a few were used in the 1975 edition. First published in 1928, this was the first of many books and magazines at home and abroad to be extensively illustrated with my photographs. There was a sequel.

I have already mentioned the Easter classes in Marine Biology for students studying zoology or botany at universities. For some years the chief naturalist, Dr J.H. Orton, conducted them as he had done when I attended in 1926. Since then I had been his assistant and was to have been so again in 1929. Suddenly he was taken ill the day before the class was due to start, and Dr Allen asked me to take over in full charge (subsequently, with the help of G.A. Steven, I ran the classes for many years). On this occasion I was asked by the students if I would take them on the Sunday for an all-day walk, so I arranged an excursion to Cawsand across Plymouth Sound, to walk around Penlee Point and Rame Head. Little did I know that the walk would be another turning point in my life.

One of the students had brought along a young woman friend, a visiting research worker in the Laboratory. As we went through the wooded path to Penlee Point this friend joined me and as we walked side by side asked if I were the D.P. Wilson whose photographs were in *The Seas*. How had I taken them? She was herself keen on photographing on the seashore, mainly seaweeds which were her specialism, and she had particularly noted the photographs which were attributed in the Preface to me. Her name was Alison Westbrook, from the Department of Botany at Westfield College in the

Alison

Rectangular tank and camera placed above

University of London; four years later she became my wife. Once again photography had led in the right direction, and her copy of *The Seas* with my photographs specially marked by her before we met is a treasured possession still.

I must return to the subject of flashlight photography. After that suggestion made to Dr Allen I bought two Kodak amateur flashpans. Each was like a cylindrical torch and contained a battery, but instead of a light bulb within a reflector the head carried a rectangular ceramic 'dish' into which flashpowder was poured. At one end of the dish a fine fuse wire glowed, igniting the powder when an electric current passed through it. I dispensed with the batteries and wired up both 'heads' to an ordinary wall electric light switch on a block of wood which I could hold in my hand. An accumulator supplied the current to fire both flashes simultaneously on a flick of the switch. A large rectangular tank with slides sloped out at 45° to avoid creating shadows was made, and painted to prevent rusting. Sands or gravels as appropriate were strewn over the bottom, or when plain backgrounds were preferred sheets of white opal or darkly coloured glass were placed on the tank bottom.

An old half-plate reflex camera on a heavy stand, adjustable for height, faced downwards over the tank. The flashpans were sited one at each end of the tank, and to shield the camera front from their light, which would have produced a reflection of the camera in the water surface, a large sheet of black paper was wrapped around the camera so that the lens peered down, as it were, through a wide black tunnel. I learnt by experience how much powder to spoon into the flashpans. With this crude arrangement the plaice were successfully photographed under water on plain backgrounds, and later various fishes and invertebrates on sandy, gravelly or rocky substrates. For small creatures pie-dishes often replaced the tank, and only one flash was needed; a large mirror propped up at the other end reflected some light into the shadows.

There were two main difficulties: firstly the flashpowder was hygroscopic and if kept exposed to the atmosphere too long absorbed moisture and would not ignite, and secondly the contacts to the fuse wire (which had to be renewed after each firing) needed to be cleaned after each ignition. Dirty contacts were a frequent cause of one or both flashes failing to fire. An occasional solution to this problem was to employ the services of a friend to thrust a lighted taper into the powder at the command 'Fire'. A much-reproduced photogaph of a hermit crab in a whelk shell with companion sea-anemone was obtained in this way. In those days such primitive methods were essential to success!

Plaice camouflaged on a gravel background

CHAPTER THREE

Fishes by Flashlight

Locally the Marine Biological Laboratory at Plymouth has been known for many years simply as 'the Aquarium', since it possesses a large and attractive aquarium to which the public is admitted - in those days for a few pennies, trawlermen free.* It was an attractive place for a photographer too, but it was only possible to get to work when it was closed to the public, in the evenings or on Sundays. And work it certainly was, for much needed to be done before a single photo could be taken.

First it was often necessary to clean the glass of the chosen tank to free it from fingermarks of eager onlookers, and the glass might have to be cleaned on the fishes' side as well. Some areas of glass were badly scratched and had to be avoided. Then all the equipment had to be carried in and set up, a job which in itself took about an hour. The camera, carrying a matt-black anti-reflection screen on its front, with a hole for the lens, was screwed to a stand where it could be raised or lowered on a vertical upright. At this period I was using a Victorian ¼-plate wooden square bellows camera called an Instanto, which belonged to the Laboratory. It had a fixed front on which the lens was mounted; the back with ground-glass screen racked to and fro for focussing. For close-up photography this is a much better arrangement than a fixed back with front focussing, as anyone who has attempted close-up work with old-time cameras will know. Clamps on tall retort-stands held the flashpans, while a heavy accumulator stood on the floor. Focussing had to be done with a black cloth over the head; the exposure, after inserting the plate, was made by removing the lens cap for a second or two while both flashes were fired simultaneously.

---

*DPW lived to know of the plans to develop a National Marine Aquarium at Sutton Pool on the Plymouth Barbican, which opened its doors in 1998. The old MBA Aquarium, vividly remembered in these pages, was closed. The area is now a Marine Life and Environmental Science Resource Centre.

Reflections had not been particularly troublesome when photographing through a water surface, but now, photographing through glass, they were. The very first trial plate showed reflections not only of the flash and an outline of the screen on the camera front, but also of the photographer with eyes closed against the blinding light! In the middle of the picture was a rather small image of a Rock Lobster.

This first attempt showed what had to be done. The flashes were moved further to each side and screened, back and sides, with tin-plate sheet to shield direct light from illuminating the camera and other objects – including the photographer, who had to move smartly out of the way. An additional and much larger black screen behind the camera supplemented that on the camera front, and the camera stand was draped with a black cloth. Once this had been done a few fairly good photographs of slow-moving fishes, and some invertebrates, were obtained.

Photographing actively moving animals was much more difficult. Solving this problem with the apparatus at my disposal and within the limits of my purse was not easy. The old Instanto camera had been much used for cap exposures; now my roller-blind shutter replaced the cap. An electro-magnet removed from an electric bell was fixed to pull up the metal

Butterfly blenny (Blennius ocellaris) *in a Bovril bottle*

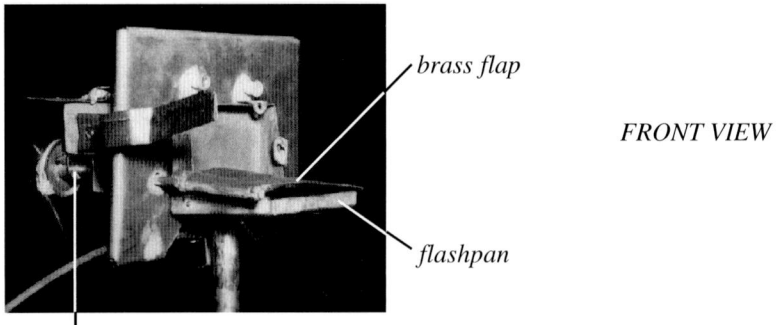

brass flap

FRONT VIEW

flashpan

meccano rod

catch (normally raised pneumatically by rubber bulb and tube) which released spring-loaded blinds. As this catch was non-magnetic a strip of iron was fastened to it. By this time I had abandoned the Kodak flashpans in favour of two Agfa ones in which the powder was ignited by a shower of sparks, produced by a toothed wheel rotating rapidly by clockwork against a flint. A long 'antinous' cable to each flashpan released the wound-up clockwork. Both were fired simultaneously when the two cables were clipped together in a single plunger held in the hand. On one pan I fitted a crude device, constructed from brass sheet and Meccano parts, in which a brass flap, blown up by explosion of the flash-powder, turned a strip of clock-spring to make violent contact with a rigid brass arm, thereby completing

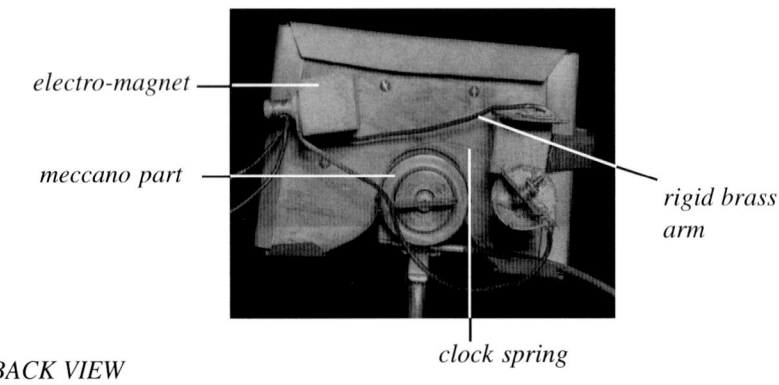

electro-magnet

meccano part

rigid brass arm

clock spring

BACK VIEW

an electric circuit to the electro-magnet activating the shutter. By this Heath Robinson arrangement the shutter blinds opened for a 45th or 90th of a second, just as the flashes reached maximum intensity. Non-hygroscopic powder was used; unfortunately it produced very much more smoke and dust than the almost smokeless powder, but that, being hygroscopic, quickly became too damp to ignite, especially in the Aquarium. A lot of powder had to be used, more than for equivalent subjects in air; sometimes it felt as if the whole place was being blown up, so fierce were the flashes, so thick the white smoke.*

But at last I was able to photograph active animals, in the Aquarium and in a photographic tank I paid a local carpenter to make. This had a base and sides of wood with plate glass windows front and back. Internally it was 18 inches long, a foot wide and 14 inches deep. Into vertical grooves inside each end could be slid glass sheets at varying distances from the front window to stop animals backing out of focus into the scenery arranged behind. Sometimes ground glass was used to avoid reflections of the far sides of fishes. Red or yellow glass backed with white opal glass gave darker or lighter-toned backgrounds. Soon I made a smaller similar tank for little creatures. The plate-glass windows were removable for cleaning; in use they were clamped against rubber strips on the front and back edges, ensuring water-tightness.

In a big aquarium tank lobsters, large crabs, anything in a settled position not too far back from the viewing window could be focussed on the ground glass screen and, with luck, photographed before they moved away. Not so swimming fishes. The camera would be focussed on a zone a little way into the tank where observation showed that the fish often swam. The field of view was marked out with small pieces of white paper stuck onto the viewing window of the tank. It then became a matter of judging when the fish was within the area of the picture, and in focus. There were many failures. After each exposure I would dash to the darkroom to develop the plate and find out what success I had had and what adjustments should

---

*Flash photography is now forbidden in public aquaria because of the stress it causes the animals. But no creature seems to have suffered from DPW's efforts.

Sea Life in Focus

Rock goby (Gobius paganellus)

be made. If at the end of a hard Sunday's work, starting after breakfast and lasting until well into the evening, six plates had been exposed and one negative was reasonably good I felt I had done well; with two good negatives, very well. Sometimes there was nothing worthwhile and one could only hope for better luck next time. It took an hour or so then to dismantle the apparatus and carry it upstairs, and to clean up at least some of the white flash dust which had settled on nearby tank sills and floor.

Early photographs in the Aquarium were invariably by flashpowder used one way or another. Later, flashbulbs and photoflood lighting were used, as described below. But for a long time a camera with a ground glass focussing screen was all I had, and for several years the photographic method was as explained. After marriage in 1931 I would often go to the Aquarium immediately after Sunday breakfast; later Alison would bring along a picnic lunch and tea. She would help me if possible, otherwise spend her time in the Laboratory library – throughout her life libraries were second homes to her.

Fishes by Flashlight

Of much concern to the aquarium photographer is the clarity of the water. At Plymouth it varied from day to day; at feeding times, every two or three days, fishes dashed about stirring up silt from the sand or gravel on the bottoms of their tanks. The silt took many hours to settle again. Sundays, when there was no feeding, were clearer days, but even so particles in suspension often obscured the backs of the larger tanks. The Aquarium had been constructed in the 1880s without any system of filtration or sedimentation. The sea water was continuously circulated by pumps from large underground reservoirs into the tanks whence it returned to the reservoirs, which were rarely emptied and refilled from the sea. In the spring of 1957 I was able to make some structural alterations whereby silt in the returning water was sedimented off and only clear water pumped back. Almost immediately rocks at the backs of the tanks, not seen properly for years, became clearly visible. And photographs improved. The Victorian aquarium tanks were broken up in 1958 and new tanks, for the design of which I was largely responsible, took their places the following year, the improved circulation system being retained. The new tanks not only gave the public better viewing but also facilitated the taking of better photographs. But I have rushed ahead and must return to earlier years.

One day in August 1929 Dr Allen asked me to inspect an enlarger he had noticed in the window of a local chemist's shop. It was a large horizontal wooden enlarger for ¼-plate glass negatives, with a massive lamp-house and cowl for some form of lighting giving off heat and requiring good ventilation, but it was without its original lamp. It was obvious that it could be fitted with an electric bulb, and so as it was in good condition I had no hesitation in buying it for the Laboratory dark room, which up to then lacked an enlarger; indeed it had very little equipment at all, apart from a few developing dishes and sundry jam-pots. Soon I found myself using this enlarger a great deal and during the next twenty or thirty years most of my enlargements were made with it. While inspecting this enlarger I was shown a second-hand Sanderson ¼-plate Field Camera, double extension, with that very desirable focussing back as well as a focussing front, and other movements. The lens was a Goerz Dagor f/6.8 in an Optimo between-lens shutter. Everything was in prime condition and all for £6. That was a lot of money in those days but I bought it for myself. When Dr Allen saw it he remarked that I would never regret the purchase and I never have. The camera did yeoman service year after year; it is now in honourable retirement. The two purchases made that day thus proved invaluable for much of my subsequent photographic work. In fact almost the first use of the enlarger was to produce a set of enlargements of the best flashlight photographs of fishes and invertebrates so far obtained, and send them to the Royal Photographic Society for possible display in their 1930 Annual Exhibition. It was an enormous thrill when the Society hung ten of them and awarded a medal for the set, something beyond my wildest dreams. In later years I was to receive their Exhibition Medal once again, and also their Rodman Medal for Photomicrography.

The newly-acquired Sanderson camera immediately took over the role of the old half-plate camera for work on the sea shore, but did not immediately displace the Instanto for tank or aquarium work. For one thing the problem of shutter synchronisation was not as easily soluble as with the roller-blind shutter, which was still working quite well. However, the solution was not long in coming. The hammer arm of an electric bell pressed on the release button of the Optimo shutter when an electric current flowed through the electro-magnet, and that took place, of course, when the circuit was completed by the brass flap arrangement on the flashpan. This synchronisation method continued in use for several years. When eventually the Agfa flashpans wore out they were replaced by two home-made flat wooden holders,

Sanderson camera with home-made adaptation

Fishes by Flashlight

Photoflood apparatus. Gummed papers mark the area in focus

covered with asbestos, on which the flashpowder was ignited by fuse wires, a return to the earliest method but with more reliable fuse wire contacts made from bulldog paper grips. One stand had, as before, a rotating brass flap to complete an electric circuit to the shutter release.

In the mid 1930s I was searching for a cleaner form of lighting. Powder flashes were particularly irritating when photographing through a water surface: some of the white dust produced settled and floated on the water and had to be removed. Large sheets of filter paper gently lowered onto the surface and carefully peeled off brought away most of the particles, but it was a tedious procedure, not completely successful. Photoflood bulbs were then coming into general photographic use and seemed to offer a solution. But it was no use to switch them on and expect the animals not instantly to dart as far away as possible. The photograph had to be taken before they had time to react to the sudden blaze of light, a matter of a fraction of a second. It also took a fraction of a second for the lamps to reach maximum illumination; the shutter had to be synchronised to make a brief exposure exactly then; if it opened too soon the lamps were not bright enough, too late and the subject had fled. Two ordinary electric light wall switches were mounted side by side on a wooden block and connected by a rod with which they were turned on or off simultaneously. One switch was on the mains circuit to the photofloods, the other on a battery circuit to an electric bell, modified so that when the hammer hit the gong it stayed there, instead of flying to and fro as in normal ringing. Through this hammer and gong a second battery circuit led to another bell similarly modified, and through the hammer and gong of the second bell a third circuit led to the shutter release. Without the fractional delay brought about by these two bells (one was insufficient) the shutter went off before the lamps had had time to light up. The photoflood lamps were in silvered glass reflectors of a type then in common use for shop window lighting. Later two sets each of three lamps were mounted side by side in reflectors made of baking tins lined with crumpled aluminium foil, giving greatly increased illumination.

About this time Sasha flashbulbs came on the market. Shaped like ordinary light bulbs they contained oxygen gas, and were packed with aluminium foil around a filament coated with an explosive paste, igniting the foil to give a brilliant flash. They were expensive to use compared with flashpowder or photoflood lighting and at first I did not make much use of them, but their cleanliness and convenience, giving a flash of shorter duration than that from flashpowder (in the quantity I had to use) told in their favour, in spite of some bulbs failing to flash. Soon afterwards more reliable types of photoflash appeared, and flashpowder was phased out.

However, it was while still using flashpowder that I had acquired a Rolleiflex, bought in the summer of 1937 primarily to photograph a lively daughter nearing her first birthday. Reflex focussing was almost ideal for such a purpose, and for lively fishes as well. But for fishes I had to combine it with flashlight. So yet another type of shutter, a Compur, had to be operated electrically when the brass-flap contact-maker on the flashpan completed the electric circuit. The Rolleiflex was mounted on a wooden box housing a mechanism made of Meccano parts, coiled springs and an electro-magnet in circuit with the device on the flashpan. When the electro-magnet released a compressed spring surrounding a sliding rod, the rod shot forwards and operated an antinous release

Hester

Fishes by Flashlight

meccano part in released position

sliding rod

to the shutter, much as one operates an antinous release between thumb and fingers. The flashpowder was fired by a fuse wire in a battery circuit completed by a foot-switch. By such means some photographs of a particularly fine shoal of mackerel in a very large tank were obtained with one-hundredth of a second exposures in the middle of the flashes, but in order to illuminate the volume of water occupied by the mackerel a really large quantity of flashpowder had to be used in each of the two flashpans, sited on either side of the photographer holding the camera and with one foot on the foot-switch. Each time the switch was pressed there was trepidation that the unusually severe heat from the flashes might shatter the glass; luckily for me it didn't!

    The outbreak of war in September 1939 brought a halt to photography in the Aquarium and on the seashore, and nearly put an end to the Aquarium itself, the Laboratory, and me too. By then the Director was Dr Stanley Kemp who succeeded Dr Allen on his retirement in 1936 – and he too was a distinguished and splendid Director. During the first months of the war there had been many air-raid alarms and a few sharp attacks on Plymouth, but no sustained blitz such as London, Coventry and other cities had had to endure. Then in March 1941 the enemy turned his attention to Plymouth and on several successive nights heavily bombed the city, causing much destruction. On the first of these nights I was with some colleagues on fire-watch duty at the Laboratory when the alarm siren went off at about 8.30pm. Grabbing our equipment and gathering in the entrance hall ready for action we soon realised that we were in for a heavy raid. Parachute flares lit up the night sky and showers of incendiary and high explosive bombs rained down on the town and all around us.

Dr Kemp, who like previous Directors lived in the house which formed the east block of the original 1888 building, took command and attempted to cheer us up by telling us a funny story. As we patrolled around the buildings to watch out for incendiary bombs a sudden shower of shrapnel, intended for the enemy, drove us inside. Almost immediately a high explosive bomb burst in our yard, just missing Dr Kemp's house. The blast threw us onto the floor of a stone-paved room which opened both into the yard and into the back of the Aquarium. We were lucky to be indoors at that moment. As we picked ourselves up we were almost immediately ankle deep in water flooding from burst tanks. A brief inspection into the public viewing-hall showed conger eels and other fish lashing about on the floor; there was nothing we could do for them. Rushing up to the upper floors of the building in a search for possible incendiaries we found broken glass everywhere. Then Dr Kemp spotted a fire in the shattered top floor of his house; this we attempted to put out with stirrup-pumps and buckets of water thrown into the flames, but to no avail.

Driven back down stairs already alight into Dr Kemp's private office, from which a door led into the main biological research laboratory with its experimental and storage tanks, we continued our efforts to douse the flames which by now had a firm hold in the house. While one man, just inside the research laboratory, worked both stirrup-pumps Dr Kemp and I within the office directed the hose nozzles onto the fire advancing towards us. Suddenly there was a noise overhead and a shout from the man at the pumps: 'Look out, the ceiling is coming down!' As I crouched down there was a roar and a huge blazing beam with a heap of red-hot ash fell between us. Picking myself free from ash part way up my legs I had to run, choking from the fumes, over the smouldering ash heap and flaming beam to reach the door to get out. Luckily for me, I had not long before got in the way of a bucket of water thrown at the flames and my clothes were wet enough not to catch fire. A steel helmet had saved my head from the shower of ash and from molten lead, though a pound or two was later found congealed on the back of my mackintosh. Only a few minor burns on my face, ears and hands had to be treated afterwards.*

Dr Kemp and I were fortunate that the beam fell between and not on us; for a dreadful moment I thought he was buried under it. He escaped down the stairs of his house, staggering out with his grandfather clock over his shoulder, and just in time to rescue his parrot. But he lost everything else: books, manuscripts and treasures of a lifetime. His house and its contents burned down; only the thick wall between it and the rest of the building saved the whole of the main Laboratory from going up in flames. A military fire-fighting party from the nearby Citadel arrived shortly after that dramatic collapse of the roof and with a powerful pump, pumping from our sea-water reservoirs, they put out the fire. Two or three days later I had the unwelcome task of recording photographically the damage done that awful night. It was years before the Aquarium was restored and tank photography could begin again.

*For many years the mackintosh, still marked from the molten lead, hung outside DPW's door in the Laboratory, the pockets used to store his keys. Now (2002) it hangs in the present Director's room.

Common Octopus (Octopus vulgaris)

Hermit crab (Eupagurus bernhardus) *with anemone hitching a ride. See p.19*

Shoal of mackerel (Scamber scombrus). *See p.29*

Young Lumpsucker (Cyclopterus lumpus)*: a sucking disc enables it to attach itself to rocks*

CHAPTER FOUR

Microscopic Monsters

'It's only a tailor's bill,' she said, as she pulled one out from an envelope I had just handed her, 'my husband is a very busy man and puts things like that into the waste-paper basket.' The husband who was so contemptuous of his bills was a director of the firm in whose Manchester office I was working in 1920. He and his wife were preparing to return to Bombay from temporary residence in Altrincham and had asked my kindly boss to send someone to collect a 'package' for shipping to India. 'Take this with you,' he had instructed me, handing over the envelope, 'and tell them that others have had to be redirected monthly – ask them to settle whatever it is.' 'Have you brought a porter?' the lady demanded next. 'No, I came on a tram.' 'Go to the station and get one, you can't carry that,' pointing to the 'package' which turned out to be a large sack containing bedding. As the porter and I returned through heavy rain to the station, with the sack on a trolley, he told me that the next train to Manchester Central Station would not be for an hour. I sat in the dank waiting room, took from my pocket a small book and began to read. I was enthralled. One of the Cambridge Manual series, it was entitled *Life in the Sea* and was written by James Johnstone of the Fisheries Laboratory, University of Liverpool. The hour passed quickly, but not before I was completely hooked on plankton, of which previously I had scarcely heard. What did I care, that memorable day, when at Central Station a porter dropped the sack into a puddle of water. The bedding would dry out before reaching India!

    I have already told how the following year I saw living plankton for the first time, under the tuition of Andrew Scott at the Piel Marine Laboratory, and how shortly afterward I was tow-netting from the pier at Rhos-on-Sea. Very soon, in my attic dark-room at home, I was photographing pickled plankton catches with the old half-plate camera supported above and connected, without the lens, to a microscope. Quite recently I came across some letters from Andrew Scott in one of which,

Microscopic Monsters

dated 20 December 1921, he pronounced as 'excellent' some prints I had sent him of those early efforts, and asked for the loan of negatives to make lantern slides to add to his collection. This gave me great encouragement, because Scott himself was a good photographer of plankton, albeit preserved as was mine. By the time I entered university I had a considerable collection of photomicrographs of pickled plankton organisms from my tow-nettings.

Plankton, an assemblage of minute plants and animals, mere specks of strange and often weird shapes, drift in untold myriads throughout the seas and oceans of the world. A fine-meshed net drawn through the surface water of the sea almost anywhere will inevitably strain out two sorts of miniscule organisms, tiny shrimp-like animals and much smaller single-celled plants. The former, copepods of many different kinds, abound in waters salt and fresh, and are among the most abundant animals on earth.

Bill Searle, veteran collector for the MBA, drift-netting for specimens from Gammarus, the Laboratory boat

Sea Life in Focus

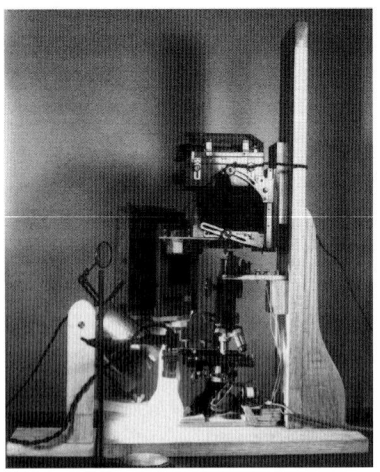

Apparatus for photographing living plankton

The latter, diatoms with siliceous skeletons, must be even more numerous, for they are a basic food of countless copepods and of many other animals capable of devouring them in quantity. Along with other tiny single-celled plants swarming in the upper layers of seas and oceans as far down as sunlight penetrates, they are the floating pastures on which almost all marine animal life is nourished. Copepods and other animals eat the minute plants and are themselves eaten by larger animals, for instance herring; these are consumed by still bigger fishes, sea mammals, sea birds and man.

Thus copepods and diatoms were prime subjects for my earliest efforts in plankton photography, but it was many years before I could take photographs of living plankton, and during those years other photographers seem not to have had any success either. Such photographs as appeared in books were of dead or motionless organisms; the technology to do better was probably not then available.

It was after the introduction of photoflood bulbs in the mid-1930s that I devised a means of using them with the Sanderson camera and my own microscope to obtain my first pictures of living plankton. In a photographic shop in Bath I saw and promptly bought for a few shillings the remains of an old half-plate field camera consisting only of ground-glass screen, back, front and bellows. Back in Plymouth this remnant was set up horizontally to one side and at the level of the microscope eyepiece. Just above the latter a right-angle prism, borrowed from my wife's microscope drawing apparatus, directed the microscope image onto the ground-glass screen of the half-plate relic, where it could be examined for sharpness with a reading lens on an adjustable arm. Vertically above the prism and eyepiece was the Sanderson camera, with the lens but not the Opimo shutter removed. There were two sources of illumination: a photoflood, to give light for the exposure, was reflected up the microscope by the microscope mirror, while over the latter a second mirror similarly reflected light for focussing from a 60 watt opal lamp in a cocoa-tin housing. The photoflood bulb was also screened and masked, and between it and the microscope mirror a bull's-eye condenser was placed to concentrate the light. Both prism and secondary mirror were carried by spring devices (modified bell-push room-indicators) which when electrically released flicked them out of the way immediately before the Optimo shutter was electrically tripped. Adjustments were made to ensure that the image when sharp on the side screen would also be sharp on the plate in the camera when that was exposed. With prism and secondary mirror in focussing position I sat in semi-darkness watching the side screen, focussing the miscroscope with one hand and moving the vessel containing the plankton with the other. As soon as an animal was in sharp focus a foot-switch was depressed, lighting up the photoflood and causing the prism and secondary mirror to swing out of the way. A fraction of a second later the Optimo shutter, set at one-hundredth of a second, exposed the plate. By these means a number of organisms were photographed successfully, but with very active animals it was exhausting work, more so than the description implies.

Close-up showing the focussing mirror flicked out of the way

One of the most successful pictures obtained using this method was of the large copepod *Calanus*, a major food of the herring. Two long bristled antennae stretch out to either side from the head end of the segmented oval body, aiding suspension while the copepod is rowed along by various appendages on the underside of the body, which ends in a segmented and

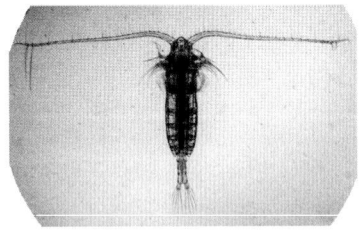

Calanus helgolandicus *photographed with the apparatus shown on previous pages*

forked tail. Slow steady forward swimming is repeatedly interrupted by sudden quick darts. Other free-swimming copepods follow the same general pattern; there is variation in size, most being much smaller than *Calanus*, some feathery with long bristles on antennae, limbs and tails. A single eye of three sensitive units is usually present in the middle front of the head; in one common copepod there are two strangely large telescopic eyes. *Calanus* sheds eggs freely into the water but in many other species eggs are carried by the female in a cluster, or in paired egg-sacs attached to the underside of her abdomen. Eggs hatch into a distinct type of larva, the nauplius, which eventually reaches adult status after a series of moults, increasing in size and complexity each time.

By the end of the 1930s I had abandoned the arduous method described above and some years passed before a much easier way was tried. It had been found that for marine animals a solution of 7.5% magnesium chloride in freshwater successfully narcotised sea-anemones and other animals put into it from sea water, preventing contraction during preservation. It was preferable to other narcotics such as cocaine, often used for that purpose in those days, and was in general use in the Laboratory. Would it work with plankton animals? It did, beyond hopeful expectations. In November 1942 I first began to use it for active plankton animals pipetted into it from sea water; most quietened down into perfect states of expansion or relaxation with

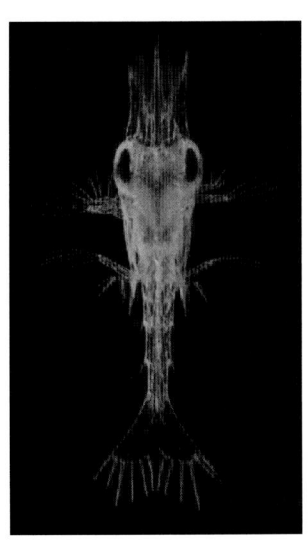

Narcotised Squat Lobster larva (Galathea strigosa)

inhibition of muscular movement. Transparency was unaffected and usually limbs and other appendages assumed natural postures. Ciliary movements – the lashing of minute hair-like structures called cilia by which many organisms swam – were unaffected, and such organisms swam as before, but without muscular contractions of gut or body wall. On return to sea water muscular movements were gradually restored and most organisms recovered completely if not kept in the solution too long. It was even possible to narcotise the same larva several times during a period of days without affecting its development compared with others not subjected to the treatment.

Now the taking of photomicrographs had become largely a matter of manipulative skill with fine pipettes and needles under the microscope, guiding an organism into position and removing any unwanted particle of extraneous matter. A ciliated organism had to be gently trapped under a scrupulously clean cover glass lowered onto it just sufficiently to prevent it from swimming away, taking care there was no distortion. For this purpose I used a Rousselet compressorium, a device well-known to Victorian microscopists but apparently no longer made, although often useful when examining living creatures. Dark ground illumination, achieved by inserting a centre stop under the lower lens of an Abbé condenser, proved on the whole to be more satisfactory for revealing details of transparent objects than direct transmitted light, and was almost always used. The special camera, an inverted aluminium cone carrying a plate-holder on the upper wide end, fitted at the lower apical end over the microscope eyepiece. Just above the eyepiece was a side tube for observation and focussing, and a shutter with an antinous cable release. The latter was clipped to a home-made device which pressed it as between finger and thumb, when a spring was released electrically via a foot-switch, thus leaving the hands free for other purposes. Exposures with an ordinary microscope lamp were in the range of one-fifth to one second, rarely longer; with narcotised animals very short exposures are not necessary. Of course the still rapidly beating cilia of many larvae were photographically blurred, but that was much closer in appearance to what the eye normally sees than the frozen beats of later electronic flash pictures.

Successful as was the narcotisation technique, the challenge to photograph plankton alive and kicking was still there. Moreover, narcotisation was not universally successful; for instance the lovely swimming organ, or velum, of some mollusc larvae was never fully expanded: such larvae needed to be photographed in a reasonable depth of water while swimming freely.

Sometime in the late 1940s we became aware of something called electronic flash, giving very bright flashes of extremely short duration and suitable for photography. Early in 1948 F.J. Warren, a Laboratory technician, obtained from the Department of Physics in the University of Bristol full instructions for building an electronic apparatus using either one or two early Mullard LSD3 flash-bulbs. The set he built for me weighed about a hundredweight, so heavy were the condensers – surplus war stock bought cheaply. The whole thing was contained in a large strong wooden box on castors. It worked well and what were almost certainly the first electronic flash photographs of living marine plankton were obtained with it, some of which were published in the popular magazine *Picture Post* on 27 January 1951.

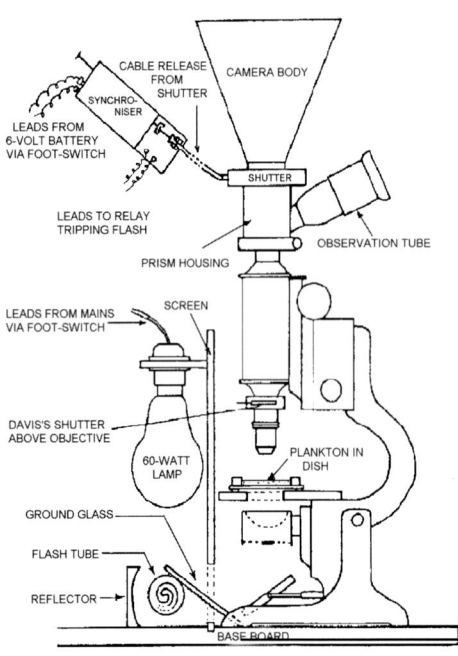

The problem posed by the need to use one source of light for viewing and focussing, and another, the flash, for photographing, was solved in a simple manner with the aid of a glass microscope slide, ground with carborundum powder on one side. The slide was fixed at 45° just in front of the flash-tube, the light from which it diffused, while the upper shiny side reflected

DPW using the apparatus shown in diagram form opposite

light from a 60 watt opal bulb directly above it. Thus there was no longer need for two mirrors. The microscope, microscope camera, side observation tube and so on were as described for narcotisation photography, but now a synchroniser was incorporated to activate the flash-tube while the shutter, set at one-tenth of a second, was open. As usual this was triggered with a foot-switch, which also put out the viewing light while the shutter was open, so avoiding ghost images. The flash had a duration of about one three-thousandth of a second, freezing all movement. The only difficulty remaining was that of depth of focus, partly solved by the use of a small iris diaphragm (Davis's shutter) above the microscope objective.

For more than a quarter of a century this method continued more or less unchanged. Only a succession of increasingly smaller and lighter, ultimately pocket-sized, electronic flash units replaced the massive and heavy set of 1948. The shutters of modern cameras themselves trigger the flash, but I still used the foot-switch and arranged for the camera shutter to be activated electrically through a cable release as before. For pictures of plankton animals too big to be photographed whole with the lowest powers of the microscope, and for pictures of typical plankton catches showing mixtures of species, a special wooden stand was constructed in which plankton in a flat-bottomed dish were lit from below by an electronic ring-flash. This gave dark ground illumination similar to that obtained on a

Sea Life in Focus

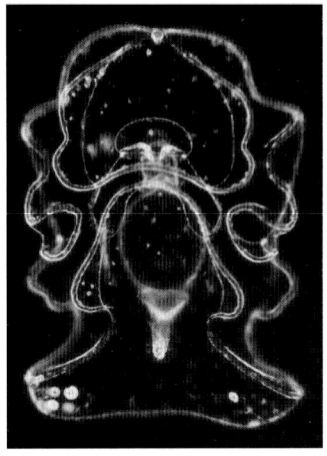

Larva of the Burrowing Sea-Cucumber with waving ciliated band

microscope, but covering a much greater area. Small electronic flashes could add light from above. Various cameras were used on the stand, fitted with lenses of short focal length, or ordinary lenses on extension tubes, which were useful also for opaque or semi-opaque small non-planktonic animals.

At first only black and white pictures could be taken, but after the introduction of Kodachrome II, with its increased speed, good colour photomicrography became possible. Many of the earlier black and white photomicrographs illustrate Sir Alister Hardy's wonderful book *The Open Sea (Part I): The World of Plankton*, first published in 1956 by Collins in their New Naturalist series. It was after this book appeared that requests for plankton pictures began to pour in from almost all over the world.

Publication of these photographs undoubtedly helped to stimulate awareness of the existence of the minute plants and animals of the plankton, but only detailed studies can fully reveal the incredible variety of their life forms. Here it is possible to refer only briefly to a few of the very many organisms which, in addition to the copepods and diatoms, abound in the marine plankton. There are the flagellates and dinoflagellates of the plant plankton for example, amongst which is *Noctiluca*, a luminous dinoflagellate. It has an animal mode of life, feeding on other tiny plants and animals

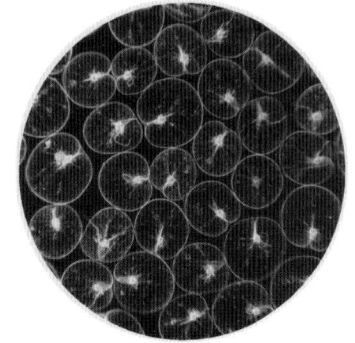

Noctiluca scintillans

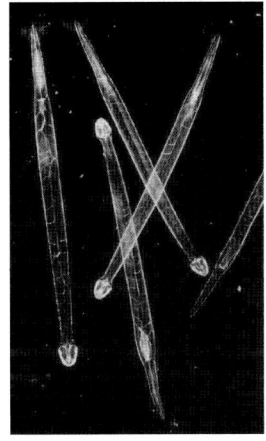

Arrow-worms from the marine plankton

instead of functioning like most plants. When it is present in dense swarms the sea may be tinged pink during daylight, but at night when agitated, as by the dip of an oar or the breaking of a wave, the sea becomes brilliantly luminescent, the water glowing with light. Of true animals, as with plants, there seem to be countless numbers differing greatly in size and form. There are tiny crustaceans such as *Podon* and *Evadne*, relatives of the freshwater *Daphnia* so well known to home aquarists as an excellent fish food. There are the glassily transparent darting arrow-worms with powerful teeth-like bristles with which to seize copepods, larval fishes and other prey. Swimming sea-snails, pteropods or sea-butterflies, row themselves with large wing-like lobes extended out of their shells. The list seems infinite, yet these are all creatures which spend their whole lives in the plankton: there are other and in some ways even more fascinating objects, the eggs and larvae of numerous bottom-living animals and pelagic fishes.

In my career as a marine biologist I have been particularly concerned with the developments of bristle-worms (*polychaeta*). A typical larva when it first swims has a band of rapidly beating hair-like cilia, ringed around the head in front of the mouth and often another ring around the anus, to propel it along. As the larva grows, its body between the rings elongates and becomes segmented, the segments

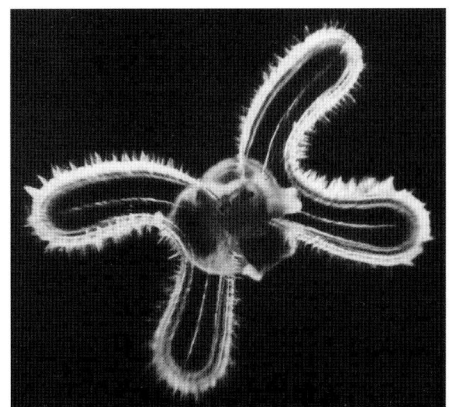

Larva of a Netted Dog-Whelk

acquiring bristles on 'feet' (parapodia) which develop on each side. Eventually the larva settles on the sea-bed, loses its cilia, and begins its adult life.

There are innumerable variations on this basic design; rarely is development as simple as that. Take, for instance, the remarkable so-called Mitraria larva of a bottom dwelling worm *Owenia fusiformis*, which I investigated in great detail in the early 1930s. In this larva the front band or ring of cilia, at first circular, enlarges into a number of graceful loops; as the segmented body forms, it folds over on itself almost like the top of a sock pulled down over the foot. The turned-over part is the front end, less the head, of the future body of the worm, and so the body wall at that end is inside-out. The head meanwhile is some distance away at the top of the larva, with the looped band of cilia in between. A striking feature is the bundle of very long protective larval bristles, arising from a little muscular cushion beside the folded-over body, which when danger threatens are spread out to point in all directions. I have seen tiny fishes which had seized larvae spit them out as soon as the bristles were raised.

When after a month or so of planktonic life the larvae are ready to settle they must make contact with a sandy bottom suited to their adult way of life. This they will spend buried in the sand, inside elastic tubes they themselves will secrete and to the outsides of which they will attach flat shell fragments selected for size and shape, fastening each fragment along one edge so as to overlap them like tiles on a roof, surrounding themselves, as it were, with a suit of scale armour. After a larva has touched down onto and identified the right kind of sandy

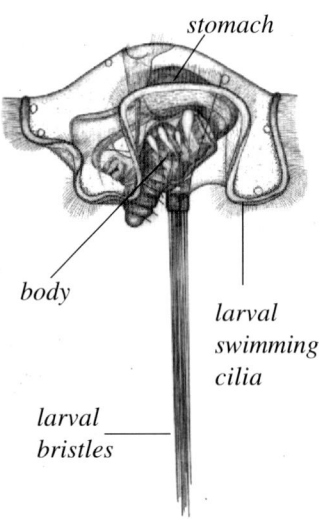

Drawing by DPW of mitraria larva 27 days old just before metamorphosis

ground an astonishing thing happens: suddenly the folded-over main body of the future worm is wriggled out into the correct posture inside-in instead of inside-out, and then connecting muscles pull down the distant head to fit onto it. The larval wall and elegantly looped ciliated band formerly between head and body crumple and disintegrate, to be swallowed by the baby worm which has already secreted a tiny tube around itself. Meanwhile the larval bristles have fallen away. The first and most fundamental changes in this amazing metamorphosis occupy seconds rather than minutes, within which space of time a beautiful swimming creature is tranformed into an insignificant-looking worm busily engaged in swallowing what had been part of itself. It never swims again.

There are other polychaete larva which have almost equally exciting developments, and among these are the larvae of the Honeycomb Worm *Sabellaria*, to be described in a later chapter. Only in the last years of studying such larvae was I able to photograph them alive; in early years they had to be drawn, accurately to scale, first in pencil on squared paper using a squared net-micrometer in the eyepiece of the microscope, finally to be traced onto Bristol board and usually inked in with Indian ink. Such drawings, it must be admitted, are often superior to photographs; not only do they force the researcher to pay close attention to details, they also enable important structures which would be scarcely visible in a photograph to be emphasised. On the other hand, photographs give a better overall general impression of appearance than do line drawings.

Some of the most beautiful of all larvae are those of the starfishes, sea-urchins and their relatives. I could never resist photographing them, from newly shed eggs or very early stages to recently settled young. All the larvae of this group have bands of cilia, often arranged in loops and patterns of great beauty. In sea-urchin and brittlestar larvae the loops form long arms supported internally by calcareous rods: all are lost at settlement, absorbed or discarded. The individual cilia of the bands are extremely slender and do not appear in the photographs; only the basal bands are visible. Equally fascinating are the larval stages of crabs and lobsters, which in spring and summer abound in the plankton, as do larvae of most sea-bottom animals. The eggs and larvae of many fishes are there too.

Sea Life in Focus

With all these wonderful forms of life in the plankton, rarely seen by anyone except marine biologists, is it surprising that my enthusiasm, stirred up in strange circumstances so very long ago, has never waned?

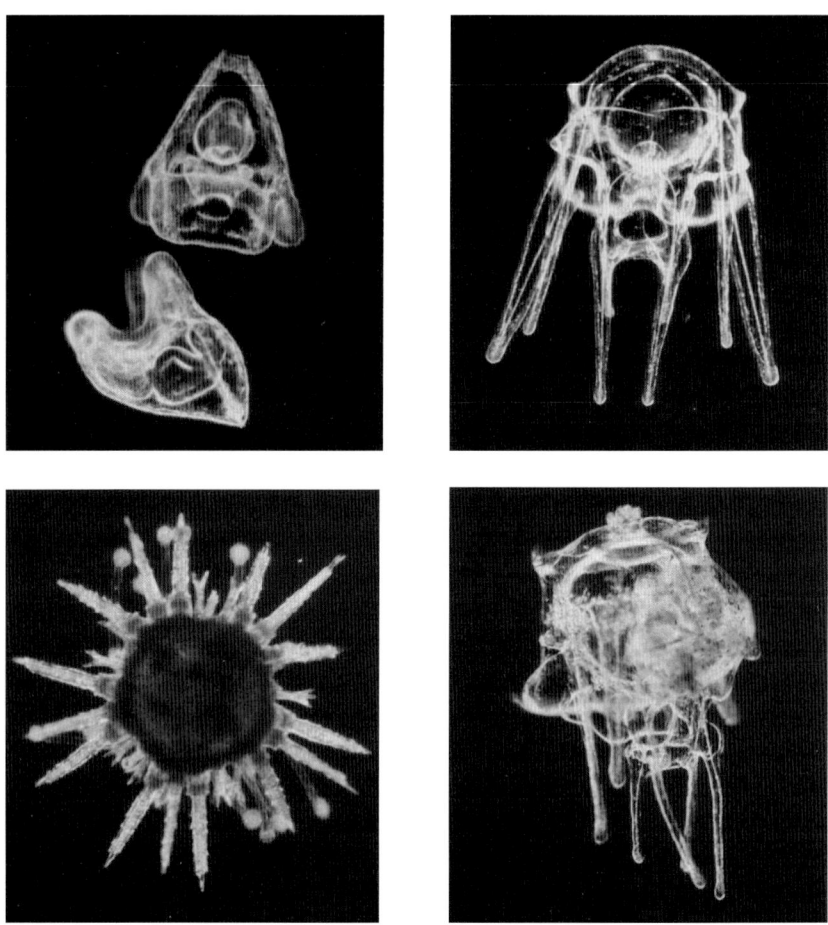

Development of Sea Urchin larva clockwise from top left: early larval stage; fully-developed larva; larva metamorphosing; the young sea-urchin

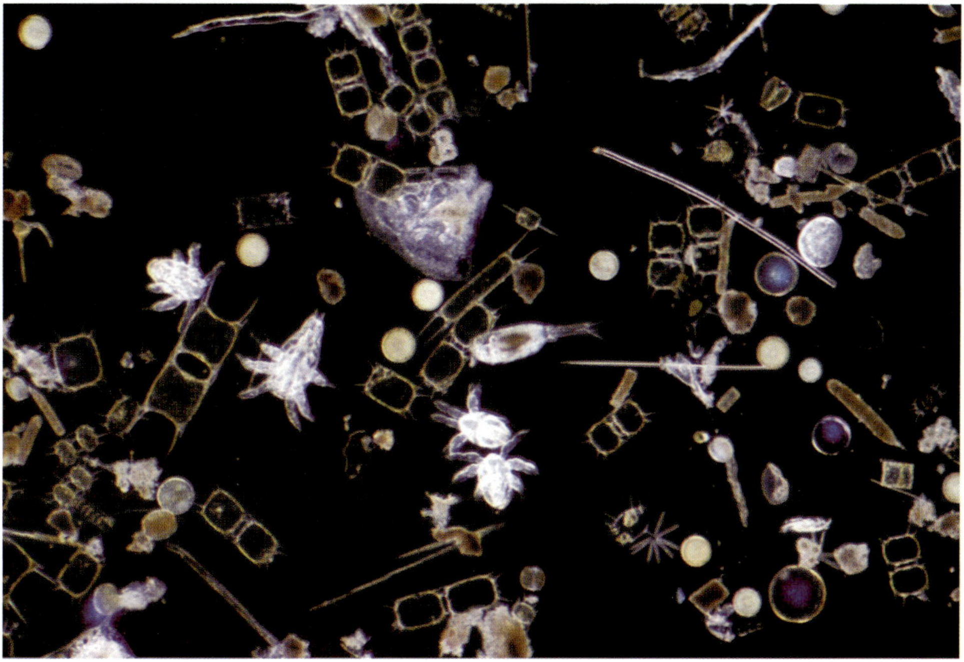

Above: *Living plankton, showing the variety of strange shapes and sizes which make up the plankton 'soup', including plant diatoms* (Phytoplankton) *and animal larvae* (Zooplankton). Below: *Green* Phytoplankton *in the spring*

Above: *The beautiful Pink Sea-fan* (Eunicella verrucosa), *a small coral found on rocky seabeds in the South West. It lives for up to 40 years, feeding on particles which float past.*

Left: *Edible Sea Urchin* (Echinus esculentus) *with extended tube feet with which it hauls itself along the rocks. In some countries the roes found at breeding time are considered a delicacy.*

Above: *Solitary Cup Corals* (Caryophyllia clavus) *are abundant in deeper waters off England, France and Spain. They can hold to the sea-bed by their pointed ends.*

Below: *The common European brittle star* (Ophiothrix fragilis) *carpets parts of the floor of the English Channel. As the name suggests, they are very fragile, and if handled may throw off all their arms at once.*

Above: *John Dory* (Zeus faber) *among sponges, sea-fans and anemones*
Below: *Cuckoo Wrasse* (Labrus ossifagus) *in breeding colours*

CHAPTER FIVE

Underwater Stories

As I have said, my interest in fishes goes back to childhood even though we lived in Manchester, far from the sea. My mother used to tell me of the old Manchester Aquarium which had been at the end of Mayfield Road where we lived, and how one day she had seen a dead shark on a trolley outside. Though it had become St Bede's College, a boys' school, one wall with fishy gargoyles was a familiar sight and helped stimulate my interest.

I was still a schoolboy myself when a next-door neighbour offered me a large iron water-lily tank, no longer in use, if I could get it from his garden to ours. It was about four and a half feet long, two feet wide and nine inches deep to the overflow hole, and was very heavy. Enlisting the services of several school friends I managed with their help to part lift, part drag it on rollers down the neighbour's garden path, out through his gate and into ours, then along our path to a prepared site against a wall shaded by pear trees. It was exhausting work and in the middle of it the back brace buttons which held my trousers up burst off and I had to retire for sewing repairs.

The tank was filled with water carried in buckets from an outside tap; in those days Manchester water was not chlorinated, was delicious to drink, and if run through a muslin bag rewarded one with a fine collection of living micro-organisms straight from Thirlmere in the Lake District. Once a lovely little *Asselus*, a sort of freshwater woodlouse, appeared all alive and kicking. I spread sand on the tank bottom, added a few rocks and provided two or three 'caves' by propping up slates along the sides. Water weeds, grass-like *Valisneria* and Canadian pond-weed *Elodia*, were planted, and duckweed *Lemna* sprinkled on the surface. The slippery filamentous dark-green alga *Spirogyra*, the curse of many a garden pond, arrived of its own accord and later on had frequently to be scooped out by hand.

After allowing the tank to settle for two or three weeks, four rudd were purchased at a town pet-shop and brought home in a large glass jar, nursed on my knees in a tram-car. The shopkeeper had at first queried my jar as unsuitable to keep fish in, but when I explained about the tank agreed to sell them. Tipped into the tank the fish immediately disappeared into the caves and it was some days before they got used to my appearing to look at them. Soon a couple of small carp were added and a three-spined stickleback. Water snails were put in to help keep down the growth of the weeds. These fish lived in the tank for several years, and occasionally others were added, including a loach. It, however, was taken one day by a heron which had discovered the tank; thereafter the tank was kept covered over with netting.

DPW and his mother in their Manchester garden

There seemed to be no way of photographing these fishes except by removing them to a glass-fronted tank. I had previously photographed fishes, but only dead ones, an eel and a plaice, bought specially from the local fishmonger to whom I had to explain that I didn't want them gutted or the heads cut off. Laid out over some stones in the garden rockery their portraits were taken (this was soon after I had bought the half-plate camera). Needless to say the results were not worth keeping – though the fish were worth eating. For the fish in my tank I constructed a make-shift glass-fronted tank but it was rather small and the poor carp which had to put up with it in the sunshine didn't like it a bit. The resulting photograph was hopeless. Good photographs of fishes had to wait until I got to Plymouth and had my sudden inspiration to use flashlight, after which I spent many hours before the war in the Aquarium with my various cameras and home-made apparatus.

After damage to the Aquarium in the air-raid of March 1941 very little photography of any kind was done for some years. Then in 1946 the Aquarium was repaired and late that year re-opened to the public. It was fortunate that most of the damage had been to the plate glass; the cast iron frames which had held the glass fronts were intact and so generally were the massive slates of backs, sides and floors. During repairs the opportunity was taken to build more natural-looking rockwork into some of the tanks before new one-inch thick plate glass windows were put in place, and to my delight the latter were as yet unscratched! But clean sea water was still ten years away; until then all photographs almost invariably suffered from suspended particles glistening in the light of the flash.

However, I did not immediately rush into the Aquarium with a camera. Just before the war I had experienced the advantage of using my Rolleiflex for swimming fishes, photographing with light from powder flashes with which high shutter speeds were easily synchronised. Now Sashalites were being used, and their shorter peak illumination made such synchronisation much more difficult. For various reasons commercial synchronisers of the time were not suitable for my purposes, making it necessary to devise and construct my own. Late in 1945 I had succeeded in synchronising the Optimo shutter on the Sanderson camera, and had obtained a series of photographs of a cuttlefish catching a prawn in one of my photographic tanks. Synchronisation of the Compur of the Rolleiflex presented other problems, especially as experience had shown that tripping it with a cable release, as in the late 1930s, was too variable in timing to catch the peak of a Sashalite flash. What was needed was a reasonably small, light-weight synchroniser which could be carried around in close contact with the Rolleiflex. After much thought, and many evenings and wet weekends spent at my garage work-bench or at the kitchen table, a complicated apparatus emerged too intricate to describe in detail here. It was my final and most sophisticated piece of gadgetry.

Suffice it to say that within a flattish open-ended 'box' of wood on top of which the Rolleiflex was mounted, there was an arrangement of levers, catches, springs, electro-magnet and several electric circuits. Power came from batteries in yet another box attached below. The whole complex could be hung on a cord around my neck, and for the first time allowed freedom of movement around the Aquarium, to take advantage of whatever might be happening in any tank.

It was towards the end of 1947 that the Rolleiflex was ready for flashlight work in the public Aquarium. In December that year four evenings were spent in front of a very big tank containing some fine cod, waiting for moments when in their apparently aimless wanderings around the tank one or more of the fish presented themselves close enough to the glass for good photographs to be possible. They most often kept to the back, wary perhaps of the shining flash reflector pointing at them. One evening I asked a watching colleague to get onto a plank walkway above the tank and coax a fish or two gently towards the front. I should have known better – that colleague never did things by halves. The almost immediate response to his well-intentioned but far too vigorous efforts with a long-handled net was for one startled cod to leap right over the top of the tank front, landing on my head and slithering down me to the floor in a lashing frenzy. We rescued it as quickly as possible but it died, poor thing, a few hours later. Four evenings of effort produced just four or five reasonably good pictures.

The Rolleiflex continued in use for several years, whenever I had an urge to photograph fishes and the larger invertebrates in the big public tanks. In later years a variety of flashbulbs more reliable than the older Sashalites were used, of short enough duration to photograph successfully all but the fastest swimmers; they were often fired while the shutter was open for a tenth of a second. Then early in 1954, through the kindness of an American biologist friend I received an Exacta Varex 35mm camera, posted to me from the Vatican City at a time when such cameras were unobtainable in this country (incidentally someone at the Post Office took the Vatican stamps which were on a document accompanying the parcel). The arrival of this camera ended the era of gadgetry, for its shutter mechanism, as in all such modern cameras, had built-in synchronisation

Underwater Stories

Cod (Gadus callarias)

for all kinds of flash. At the same time I bought a Mecablitz 500 electronic flashgun, rendering unnecessary the use of flashbulbs. The power-pack of the Mecablitz weighed only five or six pounds and could be slung over a shoulder; it had a lead to its flash-tube in a reflector which could be counted on the camera. Both Exacta and Mecablitz accompanied me later that year to the Stazione Zoologica in Naples, where I photographed Mediterranean life in an aquarium considerably older than that at Plymouth, and two years later, in 1956, they both went with me to the USA, there to be used in several big modern aquaria such as those at La Jolla, Los Angeles, San Francisco and Chicago, and huge oceanaria in California and Florida.

Through the 1950s, 1960s and well into the 1970s there followed a succession of 35mm cameras, Exactas and Pentaxes, lenses of varied focal

Sea Life in Focus

Lion-fish (Pterois volitans) *from the Pacific Ocean, photographed in California*

lengths, extension fittings and so on, and a succession of ever smaller and lighter electronic flash sets, used singly or two or three at a time, not forgetting slave flash units which could be fixed above a tank to give top lighting, flashing in synchronisation with the main flash beside the camera. Modern technology now did easily, and sometimes better, what crude home-made contrivances had achieved before, and with much less effort. Whereas in the earliest years a whole day's hard work would, with luck, produce two or three good black and white glass negatives, with this modern equipment I felt I had done badly if at the end of a lunch-hour spent in the Aquarium, even with the public present and the exposing of a whole roll of colour film, less than half-a-dozen good transparencies were obtained. The actual

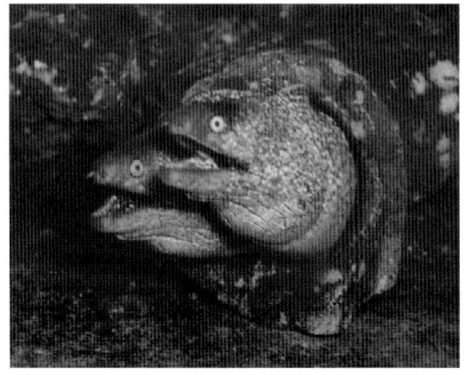

Moray Eels in a Roman amphora, Naples Aquarium

photography had become almost too easy; only the subjects themselves, the fishes and other inhabitants of the tanks, could be as unaccommodating as they always had been.

Photographing animals in aquarium tanks can have a more serious purpose than the acquiring of good pictures for personal achievement and enjoyment. The pictures eventually provided illustrations for many works of educational value published in this country and abroad. Several series of photographs recorded the behaviour, not previously observed, of a number of species, especially when powder flashes gave way to other forms of lighting enabling exposures to be made quickly one after another. In the early days of flash-powder long delays between exposures were inevitable while flash-pans were cleaned and recharged with powder. Thus in the mid-thirties when I was studying the habits of the Angler-fish (*Lophius piscatorius*) the most that could be done was to take an occasional flash-powder picture of one or two young angler-fishes lying in wait for their prey.

Angler-fishes are extremely difficult to acclimatise to captivity but I had some success with a few small ones, caught without damage, and varying from six to just over eleven inches long. These lived for several months, one for almost a year, during which periods they grew several inches. Previously there had been few concrete observations of the habits of the species, indeed almost the best account was that of Aristotle, in the third century BC, who seems to have spent much time watching fish in the clear shallow waters of the Aegean Sea. I was able to confirm much of his

Angler fish camouflaged on the bottom

account and to add to what little had been written about this fish up till then. Briefly, the angler lies perfectly still on the bottom, camouflaged to match its surroundings, until one or more small fishes are observed approaching; it then raises the modified first spine of the dorsal fin on the end of which is a tag of skin. This is jerked to and fro to entice a fish to swim up to seize it. As the fish approaches the bait it is flicked down in front of the angler's huge mouth; the victim follows it, only to be snapped up faster than the human eye can see what is happening, the next recognisable event being that of its tail disappearing through firmly closed lips. What a pity that electronic flash had not then arrived to record details of that incredibly swift capture! My anglers had distinct preferences for the kinds of fish they caught, preferring soft-finned round fishes such as pollack, whiting, mullet and gobies. They avoided hard-skinned and spiny-finned fishes such as wrasses, sea-sticklebacks and to some extent bass. They were not very keen on flat-fishes, though these were occasionally swallowed, while awkwardly-shaped species such as small gurnards and pipe-fishes were always refused. Gurnards have, however, been recorded by other observers in the stomachs of large angler-fishes.

Another method of capture which takes place too quickly for the eye to follow was recorded photographically ten years later when techniques had improved. Cuttlefish (*Sepia officinalis*) were regular inhabitants of a large tank where they were fed on prawns and small crabs. The cuttlefish has eight shortish arms with suckers, and two long ones which are normally folded up and out of sight among the others. These long arms or tentacles have enlarged extremities with powerful suckers for holding prey. The cuttlefish swims gently by undulations of fins bordering the body, and by squirts of water (jet-propulsion) from a tube, the siphon, close under the head. The siphon can be turned to direct the jet of water in almost any direction, even backwards, and the more forceful the jets the more rapid the movement in the opposite direction. The cuttlefish uses this manoeuvrability to great effect when stalking prey, or when retreating rapidly from danger, then often shooting out a cloud of ink to confuse an attacker. While hunting it glides slowly towards the crab or prawn it intends to capture; a crab aware of the danger faces the enemy, claws at the ready, turning as the cuttlefish tries to seize it from behind. When a crab does manage to use its claws the cuttlefish often lets go and the crab escapes, but if it is taken from behind it hasn't a chance. Prawns are more vulnerable and are seized from the side. As the cuttlefish advances upon its prey the two long tentacles with their grasping suckers are slowly protruded a little from between the eight shorter ones, then shot out at full length. The luckless victim is pulled back to the mouth between the bases of the short tentacles, now opened out to receive and hold it. The beak within the mouth bites into the back of the captive which quickly dies, poisoned by salivary juices.

To obtain photographs of the capture of a prawn it was necessary in those days, before the arrival of reflex focussing combined with synchronised flash, to move a small cuttlefish to a tank where it would perform within an area pre-focussed with the Sanderson camera. The shutter, set at one hundredth of a second, was synchronised with Baby Sashalite bulbs. The photographs so obtained in 1945 revealed details not before fully appreciated. For instance, the prawn sometimes escaped by dodging at the right instant; then the tentacles shot out towards it hit the ground, to be thrown into loops as the cuttlefish darted forward in the foiled attempt to secure its prey.

In 1950 a Dutch zoologist, Piet Sevenster, came to Plymouth to study the breeding behaviour of the Fifteen-spined Sea Stickleback (*Spinachia spinachia*), a little fish of inshore waters. He watched his

A cuttlefish throws out its tentacles to catch a prawn

Sea Life in Focus

The prawn escapes!

sticklebacks in a large tank in a room on the same floor as mine, enabling me, when invited to photograph their nesting behaviour, to wheel along on its castors that very heavy electronic set of 1948 which I had never been able to carry down to the public Aquarium. The camera was the Sanderson on a stand placed in front of the tank; two electronic LSD3 flash-tubes in reflectors were positioned over the tank just above water level. With this combination a fascinating series of photographs was obtained.

The male sea stickleback at breeding time builds a nest in the branches of larger seaweeds, binding branches together with a white elastic thread excreted through the kidney duct. He then stuffs into the bound mass tufts of softer seaweeds which he tears from off nearby rocks, tightly binding them into the nest by passing around and around it, and wriggling through it again and again, secreting his thread as he goes. When the nest is ready he awaits the chance arrival of a female, meanwhile fiercely driving off all intruders, rival males and even females, which depart at once if not ready to spawn. A female ready to spawn however does not retreat even if bitten; she turns towards him and he recognises this as a signal that she will enter his nest. His next move is to show her the nest, which he does by repeatedly jerking his snout into its opening. She soon responds by inserting her own snout and wriggles hard to get in – harder for her than for him because she is swollen with eggs. He encourages her by gently biting her tail. When about halfway through she spawns, the male entering beside her to shed his milt and fertilise the eggs. He comes out first and when the female follows he ungallantly attacks her and she flees the nest, having nothing further to do with the eggs. The male may now succeed in inducing other females to add to the eggs already in the nest. The eggs take about three weeks to hatch, during which time the male stands guard, not merely passively but actively, constantly fanning water with his pectoral fins onto and into the nest, to ensure a sufficient supply of oxygen to the developing eggs. When the young hatch he keeps them in the nest as long as he can until eventually they swarm out and begin to hunt for tiny creatures to eat. When first hatched they lack the long snout of the adult; this develops as they grow. Many fascinating photographs of this nesting behaviour were obtained, some of them eventually published in *Picture Post* and elsewhere.

One of the most handsome fishes ever to grace the Plymouth Aquarium was found inside a tea-chest floating on its side five miles off the South Devon coast in September 1951. Brought in by the fisherman

The male Sea Stickleback encourages the female to enter his nest

Wreck-fish

who found it, it lived in one of our large tanks for many years. It was a Wreck-fish or Stone-bass (*Polyprion americanum*), a species attracted to floating wreckage, doubtless to feed on smaller fishes sheltering and finding their food amid the seaweeds, stalked barnacles and other organisms growing on the wreckage. We fed it on freshly trawled fish such as whiting, and because of its fierceness to other large fishes (it killed a ballan wrasse about as big as itself) it had the tank to itself apart from large crabs and some smaller active fishes which for some reason it tolerated. When a second wreck-fish arrived several years later it had to be kept apart from the first, living in one of the large covered sea-water reservoirs until after the death of the first comer. Both were favourite subjects for my camera, and both attracted much attention from the visiting public.

The primary purpose of the Aquarium when it first came into use in 1888 was to enable the scientific staff of the Laboratory to study the habits, particularly breeding habits, of commercial food fishes. At that time relatively little was known and it had become essential to learn more to ensure that legislation regulating fishing and conservation of fish populations could be soundly based. Although many observations of fish behaviour were soon being made it proved more difficult than originally anticipated to induce the larger commercial species to breed in tanks; such breeding as did occur was mainly by fishes not to be seen on fishmongers' slabs. One of these was the Dragonet (*Callionymus lyra*) which formed the subject of a long and detailed account by E.W.L.Holt, published in 1898. Many years later I was able not only to photograph in colour the sexual display and mating of this fish as described by him, but also to describe and photograph the territorial behaviour of the males, which he did not see.

The Dragonet is a common fish of off-shore trawling grounds, several inches long when adult. It spends much time resting on the bottom on a flattened belly, well camouflaged by the mottled sandy colour of its back. Large pectoral fins row it skimming over the bottom in search of food: small crustaceans, molluscs and worms. February and March is the main breeding season, when exciting courtship displays take place. It is then that the male comes into full breeding dress, a brown-yellow body spotted with azure blue, pelvic fins darker blue, eyes a brilliant blue-green. His tall dorsal fin, normally collapsed along the back concealing brilliant blue and yellow bands, will at the sight of the sombre-coloured female suddenly be raised to flaunt the vivid colours as he half-circles around her whilst she lies at rest. At the same time he purses out his lips, stretching skin folds to reveal more blue and yellow bands. Should the female be tempted by this gorgeous display she moves towards him onto a fanned-out pectoral fin and together, his dorsal fin collapsed, they swim side by side steeply upwards, propelled by beats of the pectorals. As they go they bring their ventral fins into

Dragonets in nuptial display

contact and immediately begin to shed eggs and milt, continuing to do so as up and up they swim, until the last eggs are shed far above the bottom. Then they part and at once dive down again, leaving the developing eggs to drift and hatch in the plankton.

Nowadays we are well aware that many birds and mammals acquire and defend territory. Some fishes do so too and the dragonet is one of them. This became apparent when several males and females were kept in a much larger tank than that in which Holt made his observations. In the breeding season the males established territories in which they displayed to entering females, attacking encroaching males. The more pugnacious the male the larger his territory. Attacks were mostly by display against rivals, without physical contact; only occasionally were there incredibly swift encounters too quick for the eye to follow, but caught by the electronic flash. Some bites inflicted minor damage on an opponent; none was fatal.

Two more examples of the many behavioural patterns of which I have had the good fortune to be first witness and to record photographically may be summarised here. The first concerns the breeding of the Black Seabream or Old Wife (*Spondylosum cantharis*), a species which I had not seen before 1951 when a few fishes were installed in the largest tank. In succeeding years a few more were obtained and they bred regularly in springtime. When not breeding, males and females were silvery, pale violet on back and upper sides with broken horizontal stripes. This colouration seems to belie the name 'Black': it was in the excitements of nest-making, fighting, mating and guarding the eggs that the males at times changed instantly from their normal pale appearance to one of intensely dark violet-black with a prominent vertical white stripe down the middle of each side. The 'nest' was merely a circular area of slate floor of the tank cleared of its covering of small pebbles by the male swimming close above the bottom, sweeping them away with vigorous side-to-side lashings of his tail fin. He remained upright, head towards the centre of the patch being cleared, travelling around its edge with pauses for rest. Early in the season before the fish were fully mature, the nests were quite small and readily abandoned; later they reached diameters of about a metre. Males making nests side by side threatened and chased one another, sometimes facing mouth to mouth,

Male Black Sea-bream, wanting to attack a ray which is lying on its nest, is itself chased by a pout

fins erect and bodies flushed almost black with vivid white stripes. However, there was very little actual fighting.

With the nest prepared its owner would dart after any passing female until one, ready to spawn, accompanied him down to it, whereupon she tested it with her body and mouth, while the male in black and white dress, dorsal fin erect, fussed around her, nuzzling her ventral fin and anal region. Only once did I see a female change from normal pale colour. On this occasion she darkened, though less so than the male, and a pale horizontal patch appeared on her side, contrasting sharply with the vertical white stripe of her mate. The latter became wildly excited. Eggs were never laid when I was watching, but certainly they were laid, perhaps in late evening or early morning, and deposited mostly in a single layer in the middle of the patch which functioned as a nest. The male then had to keep them clean, fanning with his tail to disperse silt, sometimes removing stray pebbles in his mouth. He also had to guard them: a formidable task with so many other fishes and invertebrates in the same tank.

For nine days until they hatched the male had no rest by day, though what happened at night is uncertain. Plaice or rays straying onto the cleared patch, perhaps lying on top of the eggs, were savagely bitten until they moved away. Small rock lobsters were seized by an antenna and towed away, large ones were pushed off by the angry male applying his mouth to the back of their abdomens, then swimming vigorously. One inhabitant of the tank, however, would have none of this nonsense. A large pout, in spite of being bitten many times, often turned the tables on the guarding fish and chased him away. It did this two years running and it was all very amusing to watch. Only a sting-ray ever seemed interested in feasting on the eggs; we had to remove it to another tank.

The Cuckoo Wrasse (*Labrus mixtus*) is another of those species in which males and females are strikingly different, the male predominantly blue and orange, the female overall pink with black and white patches behind the dorsal fin. Indeed at one time males and females were considered to be separate species; recently it has been discovered that some individuals do actually change sex in the course of their lives. Breeding had never been observed until 1955 when it took place in a Plymouth tank where these off-shore rock-haunting fishes had often been kept. In that year two males and three females had been sharing the same tank without trouble until one male was found dead, apparently killed by the other which had become sexually active and engaged in nest-making. As with the black sea-bream the nest was no more than an area of tank floor cleared of gravel, with the difference that it was very much smaller, only inches across, and was produced by the fish turning over on his side while flapping vigorously with his tail, driving the gravel aside. Several such nests were made during a period of about two weeks.

Having cleared a nest patch the male turned his attention to the females, chivvying each in turn, sometimes biting them. At such times his colours were intense, as were those of the females. Unfortunately the latter were not quite ready to spawn; nevertheless one or other would eventually follow him down to his cleared patch. At this he became very excited, spreading all his fins, showing off all the magnificent colours. With open mouth he jerked his head and shoulders from side to side while, most amazing of all, there appeared on top of them a large area of skin blanched almost white, leaving only faint traces of normal stripes. His usual appearance was completely transformed. This whitish patch had previously appeared briefly, less fully blanched, during nest-making and attacks on females. These activities took place only in the evening; the white patch was never seen at other times of the day. Sadly the females never fully matured and no eggs were laid.

There is a sequel to this story. One day Alison and I were shopping in Italy when we noticed a set of postage stamps on sale. They were stamps of San Marino showing a variety of marine animals, among them a male cuckoo wrasse in sexual display with that unmistakable white patch, never recorded until I saw and photographed it. The picture had been copied from one of my photographs which, not long before, had appeared in an Italian encyclopaedia. On the stamp the picture was reversed left to right, otherwise it was almost exactly the same, even to the rock background. We bought two sets. I felt too flattered and amused to take any action against the San Marino postal authorities for this shocking infringement of my copyright!

Cuckoo Wrasse with white breeding patch

CHAPTER SIX

On the Seashore

While seashore photography required the invention and use of very little home-made apparatus, it did occupy during some sixty years innumerable hours spent wandering over many different shores in the search for pictures. But the shore where we spent most time was Trevone, near Padstow on the North Cornish coast. There, next to a fine sandy bay, is another called Newtrain where a great rocky outcrop thrusts into the sea, and this provided ideal opportunities for shore photography. The main need is, of course, to find good subjects in the time available while the tide is out, a time which can be very short at the lowest levels, which are uncovered only by spring tides. It is there that the greatest variety of life is to be found, and unless there is previous knowledge of a locality much precious time can be spent searching around for worthwhile subjects. Our familiarity with the Trevone rocks and pools was an aid to photography.

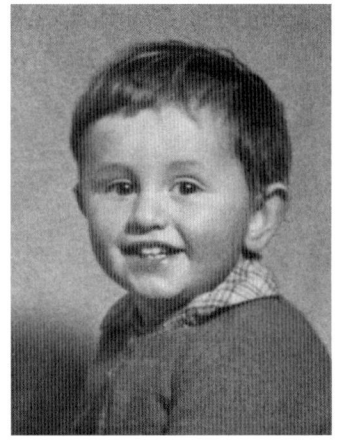

Richard

Alison and I first explored Trevone in the 1930s, initially staying in a boarding house, but later in Sandpits, a wooden chalet-caravan which nestled below sand-dunes at the back of the beach. It was lent to us by Dorothy Sewart, whom we first met on the cliffs while bird-watching and who became a lifelong friend. Tucked away behind a tamarisk hedge and with only a wall to jump down to get on to the sands, Sandpits was in a wonderful position for holidays in the late 1930s with our small daughter. Then during the winter of 1941-42 it became a refuge for Alison and Hester from the Plymouth blitz. Alison was

Newtrain Beach, looking towards Trevose Head

expecting a second baby (our son Richard). It was bitterly cold, the sea freezing on the beach, and the primitive conditions – no running water or electricity, inadequate heating and a hut with a chemical toilet – made for uncomfortable living, but it was better than bombs.

After the war Dorothy had to remove the chalet which the council thought an eye-sore (the sand-dunes were later flattened and the site used as a beach car-park). But she kindly invited us to spend August family holidays in her beautiful cottage at Lower Treneague, in a valley near Wadebridge, while she stayed in her new summer home, old fish-cellars at Port Quin which she rented for £5 a year! Lower Treneague in the late 1940s was also without main services, but pumping water daily from the well and going to bed with candles or oil-lights all added to the fun when on holiday. From the cottage we made expeditions all over Cornwall, but on the days of big spring tides we almost always headed for Trevone and its rock reef for shore photography.

Sea Life in Focus

Bladderwrack

Though such photography does not require specialised gadgetry, it does present the problems of incoming tides, weather and light conditions, all beyond the control of the photographer. We found that rockpool photography was best undertaken when there was unobscured sunshine and no wind to ripple the water surface, though on a dull day flash could be used. Care was needed to avoid reflections of white clouds, also reflections of camera or tripod, or even the photographer. A dark umbrella held vertically over the camera is said to be useful in preventing reflections, but we never used one, our cameras being directed at angles to avoid reflection. I have tried a polaroid filter, but the increased exposure time, and the need to position the camera at a rather low angle to the water surface, made it of little use. In bright sunshine wet seaweeds give many highlights from numerous reflections of the sun; it is often better to wait until they are dry, photographing before they begin to shrivel. A hazy sunny day with no wind is ideal for seaweeds, especially for the great kelps of the lower levels. But one can wait for years for all these conditions to coincide with low water of the lowest tides, and at a time when one happens to be on the shore with a camera!

At first these photographs, and those taken of animals in the aquarium tanks, were for personal enjoyment only, or for illustrating lectures to natural history societies; later more and more began to be reproduced in books and magazines, helping greatly in financing yet more photographs. My book *Life of the Shore and Shallow Sea*, lavishly illustrated in black and white, was published by Nicholson and Watson in 1935; a second edition, revised and in a new format and largely re-illustrated with later photographs, appeared in 1951. In between those years my picture book *They Live in the Sea*, illustrated with a different set of black and white photographs, was published by Collins in 1947.

By this time the use of colour films and colour printing led to an ever-increasing demand from authors and publishers for colour photographs of marine life. One special commission came in 1946, when I was invited to provide illustrations for *The Sea Shore*, written by my friend and sometime colleague Professor C.M. Yonge, and published by Collins in their now famous *New Naturalist* series. The emphasis was to be on colour

Sea Life in Focus

Photo: Alison Wilson

A natural group of top shells

photographs taken specially for the book. This was before colour films were readily available here, but it was possible to import ¼-plate sheet Kodachrome from the United States and supply it to photographers working for the series. My Sanderson field-camera was used throughout and it was a rewarding experience to take colour photographs for the first time on the seashore. Sixty of these colour photographs appeared in the book, with a similar number of mostly older black and white ones, including some by Alison. Soon afterwards 35mm Kodachrome was sold and processed in this country and at long last we were freely able to record the life of the seashore in colour, at first using a Rolleiflex, later Exacta and Pentax cameras.

Returning to familiar shore territory like Trevone can be rewarding in the way that it enables observation and recording of any dramatic changes which have taken place, or to note the development of a particular marine organism over a period of time. Duckpool near Bude, was visited over and over again to observe and photograph the growth of colonies of the so-called Honeycomb Worm (*Sabellaria alveolata*). I had been interested in it since the early days at the Laboratory, when I had reared, drawn and described the larvae of this particular worm, and of another closely-related species (*Sabellaria spinulosa*). On a number of our south-western and western shores, including those of Wales, hundreds, thousands of the *alveolata* worms live together in colonies, each worm in a tube of sand grains which it collects and cements together. The tubes are fastened side by side to produce massive reefs cemented firmly to the rocks. The colonies live only on lower shore levels where rocks and clean sand are constantly exposed to rough seas; the worms depend on the waves to wash up to them their building material, for they cannot crawl down to the sand to gather it for themselves. When submerged by the tide each worm spreads tentacles at the mouth of its tube to catch sand grains drifting by, and also to seize microscopic organisms for food. The *spinulosa* species builds similar colonies around the North Sea coasts, usually below low water mark. Off south-western coasts it is common off-shore but does not build large colonies, living singly, or a few together, in sandy tubes fastened to rocks or stones.

The larvae of the two species are very similar in their anatomy and development, spending weeks swimming in the plankton before settling down. It was long a puzzle as to how they managed to find the specialised habitats suitable for their respective adult lives, and how they avoided settling on top of one another. So in the early 1960s I began a long series of experiments which eventually led to the discovery that fully-developed larvae, when ready to settle, crawl over and explore any hard surface with which they come into contact, and that they have the astonishing ability to detect, apparently by some form of contact 'tasting', the particular kind of cement which the adult worms secrete to stick together the sand grains of their tubes. Not only can the larvae distinguish this cement from all other substances, they can distinguish it from the cement of their close relatives. Moreover, larvae can remain in a state ready to settle for weeks if necessary, giving them a reasonable chance of coming into contact with their adult habitat, as they drift along the bottom in the sea currents, testing substrates as they go. Once they find the right kind of cement they settle, throw off their protective spines, change their shape, and secrete their first transparent tube which they fix to the adult tubes of a colony, or to a rock where traces of a former colony still persist.

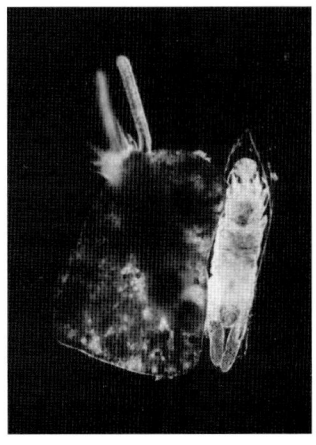

Sand grain on which two Sabellaria larvae have settled and secreted tubes

Sea Life in Focus

Alison and a huge colony of Honeycomb Worms at Duckpool near Bude, 1963

This habit of settling inevitably ensures that larvae settle always, or almost always, in the right environment for adult life. Sometimes they do establish colonies where none existed before, but usually they settle in vast numbers on colonies already established, or on rocks bearing traces of former ones. Such traces can exist for a long time. No-one knew how quickly colonies grow, how long individual worms in them survive, how long a colony as a whole exists, and what would bring about its eventual destruction. So in order to obtain answers, Duckpool, a shore where large colonies are abundant, was visited during the lowest spring tide almost every month for fifteen years, whatever the weather.

Duckpool is a sandy bay facing west to the Atlantic; out of season it is a wild and lonely place with a high rocky cliff on the north side and rugged rocks down to low water mark on the south. A wide clean sandy shore lies in between. Above high water mark a large pool, fed by a stream, overflows onto the beach; after heavy rain it is difficult, sometimes impossible to wade across. (We have seen wild ducks on that pool – ancient wildfowlers may have given the place its name for it is so called on maps as early as the seventeenth century.) The sea here is rarely calm; huge rollers from the Atlantic sweep into the bay crashing on rocks and stirring up sand, creating ideal situations on low level rocks for the Honeycomb Worm, but making it a dangerous beach for bathers. In winter Alison and I were often the only human beings on the shore and there were occasions when it would have been foolhardy to be there alone. On one such stormy day Alison was inspecting a colony when she was caught unexpectedly in the powerful swash of a huge wave which submerged her to the waist and almost pulled her under. She was soaked to the skin in a winter sea but was lucky not to be dashed against the rocks.

A number of the colonies were selected for regular photographic recording from as near standard positions as possible. Pictures were taken of each colony from front and side, with close-ups of portions, almost always with a ruler resting on the colony. From the 35mm colour transparencies enlarged black and white prints were made; the image of the ruler was cut out and used for measuring the diameters of the tube-openings. In this way many more measurements could be made than was possible on the shore within the very limited period of low tide. Over the years hundreds of colour transparencies were obtained, yielding a great amount of information, much more than could have been gained without the use of a camera. I found, for example, that individual worms can live for several years, ages up to five years being common, with some surviving for ten years or more. Deaths are generally due to the breaking up of colonies in storms, especially when attachments to the rocks have been weakened by undermining by crabs and so on. Colonies grow continually, increasing their surface areas and adding to their weights, until there comes a time when a particularly large wave crashing down on an overgrown colony breaks it up, partially or completely. Floating wreckage driven by storm waves may also be responsible for damage. Sometimes young worms will if sufficiently numerous repair a colony and give it a new lease of life. Or before they have time to do so they may perish with its final destruction.

The main spawning season in north Cornwall is a short period in July when vast numbers of eggs and sperm are shed into the sea to develop into larvae. They do so at different rates; some are ready to settle when six weeks old, others not for several months. This must help to ensure that at least some larvae from each spawning will arrive in the adult environment and recognise it as such. Newly settled young, in numbers varying from year to year, can be found on old colonies from about the end of August until the following spring, the heaviest settlements taking place in the autumn. I photographed one colony from insignificant beginnings in September 1962, through a whole series of settlements which built it up into a huge and heavy bracket on the sloping side of a rock, until most of it broke away in the autumn of 1975, leaving only a battered remnant still clinging to the rock. The life histories of worms can be quite as fascinating to study as those of many a feathery or hairy creature higher up the scale of being!

As a postscript to this story of the Honeycomb Worm I should add that the minute larvae of many other species of worms, and those of other major groups including barnacles and molluscs, also possess this astonishing ability to recognise specific habitats suited to their adult lives, an ability which is highly developed when the habitat is a specialised one of limited area. For example, the larvae of the intertidal burrowing worm *Ophelia bicornis* explore various types of sand before recognising and settling down into the particular type in which they will subsequently live. Recognition depends on the relative abundance adhering to the grains of the micro-organisms on which they will feed as adults. Too many, too few or of the wrong kind, and the sand is rejected. The 'right' sand is found only in some tidal estuaries, as at Exmouth, where I studied this species in the 1950s.

As the years went by some earlier photographs acquired unexpected interests, particularly general views of shores revealing changes, or, in some instances, the absence of substantial change. Thus photographs of the boulder-strewn shore at Penrhyn Bay, North Wales, refuted a suggestion made in 1946 by a geologist that the shore level was being washed away by storms. Although storm waves had been responsible for erosion of the unprotected cliffs at high water and damage to the promenade wall, I knew that the almost level mid-tidal shore was scarcely changed since my boyhood wanderings over it. Photographs aided memory, and one view of the shore from the promenade taken in 1946 was so well matched with another of 1974 that they must have been taken within a few inches of the same viewpoint. They produced quite a good stereoscopic image when placed side by side in a stereoscope. Several large stones and small boulders, small enough to be moved though not lifted by human effort, had retained their relative positions for at least thirty to forty years.

It was quite the reverse at Salcombe in South Devon where again comparisons of photographs many years apart proved to be invaluable in revealing the changes that had taken place. When in the late 1920s I visited the Salcombe estuary at times of low water spring tides, the exposed muddy sand banks were at the lower levels bright green, so thickly were they covered with eel-grass (*Zostera marina*). This is a true flowering plant, not a seaweed, whose roots and rhizomes bound the sand, hindering it from being washed away by the rise and fall of the tide and strong tidal currents. In the enclosed estuary, wave action is minor compared with the open coast.

Easter class students digging on the Zostera *bed at Salcombe in 1926*

Sea Life in Focus

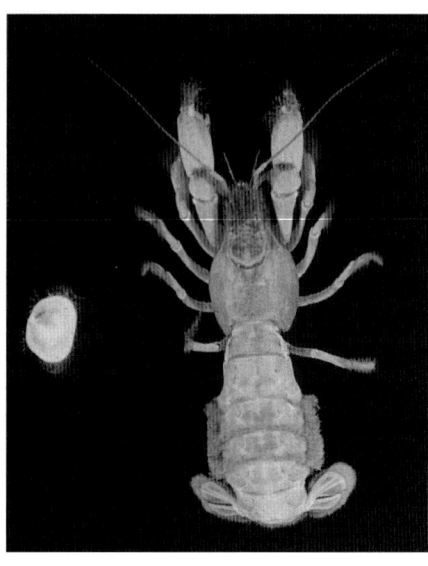

A burrowing prawn and a bivalve mollusc, which lives in its burrow

Because the plants trapped the sand – much as does the marram grass of sand dunes – the *Zostera* beds were raised above the surrounding areas, often with a steep slope at their seaward edges. In the beds there lived a rich fauna of burrowing animals, worms, molluscs, crustaceans and echinoderms. A surprising number of unrelated animals shared burrows made by larger partners. Small worms with larger worms of different species, tiny bivalves with burrowing crustaceans, burrowing brittle-stars or sea-urchins. No harm was done to the smaller creatures by their larger partners, and the smaller animals obtained shelter and no doubt food as well in this unusual house-keeping arrangement.

Examination of the so-called *commensal fauna* was a highlight of the classes for zoology students held at Plymouth each Easter vacation. With forks and spades we arrived by coach at Salcombe to dig in the *Zostera* beds for specimens, taking the animals back to the Laboratory for further study. Then early in the 1930s a mysterious disease, cause unknown, attacked the eel-grass, first on the Atlantic coasts of North America, a little later along the Atlantic coasts of Europe. At Salcombe the presence of the disease was first noticed in the spring of 1932; by the summer blackened and rotting leaves were usual. The following year the beds were almost bare of green leaves and had an overall blackish appearance. Later still with the decay of the dead roots and rhizomes the sand was loosened and began to wash away, lowering the levels of the banks by as much as two feet or more. This was shown clearly by a comparison of photographs

A burrowing brittle-star which shares its home with two tiny guests, a bivalve (outlined top) and a scale-worm (left)

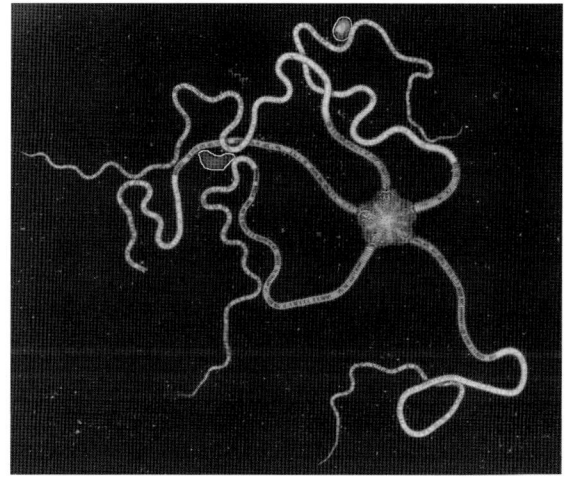

taken before the disease and afterwards. Whole areas became stony. With the loss of the sand the previously abundant burrowing animals became scarce. The washed-away sand accumulated around high water mark at the heads of several small bays on the east side of the estuary, there raising beach levels considerably, again confirmed by photographs. Some recovery of the *Zostera* eventually took place but the plants now present have, when fully grown in summer, leaves only about half as long and half as narrow as those of the 1920s, which reached lengths of almost two metres and widths of about 10mm. Those who never saw the Salcombe estuary in the 1920s can have little idea of its appearance at low water then. Large areas which were as green as a grassy field are now only areas of blackish sand or mud; such growths of *Zostera* as survive are but poor substitutes for those of the past.

On one occasion a man-made catastrophe brought about a dramatic change. On 18 March 1967 the oil-tanker Torrey Canyon grounded on the Seven Stones reef fifteen miles west of Land's End, while trying to take a short cut to catch the tide at Milford Haven. During succeeding days vast quantities of crude Kuwait oil spilled from her into the sea, much of it to come ashore on the south-west coasts of Cornwall and the north coast of

Sea Life in Focus

Hosing the rocks at Trevone in April 1967

Brittany, driven this way and that by shifts of wind direction. On 29 and 30 March much oil came ashore at Trevone Bay, coating rocks and sand thickly to more than half-an-inch deep in places. Elsewhere on the coast efforts were already in progress to clear up the mess, mostly by spraying the oil with fresh water to which had been added a detergent of a kind used by the Royal Navy to treat small oil spillages in docks. Unfortunately this detergent is highly toxic to marine life, more so than the oil itself and much more so than detergents used in such emergencies since.

At Trevone hosing began on 6 April and continued for several days with, as elsewhere, disastrous consequences for most of the fauna and flora. In particular, enormous numbers of herbivorous molluscs, limpets and top-shells were slaughtered, along with carnivorous molluscs, other kinds of animals and many seaweeds. Dead, dying, and empty shells, with a few crabs, worms and fishes, lay in drifts along the rocky gullies. By July the beach was clean for the holiday makers, but with the toxic detergent gone the rocks were bright green with dense coverings of green algae, grass-like *Enteromorpha* and the Sea Lettuce *Ulva*. Normally these green seaweeds cover only small areas of rock, or live in rockpools, kept in check by browsing limpets and top-shells. By the autumn, however, the greens were being suppressed by the browns, as young plants of Bladder Wrack (*Fucus vesiculosus*) and Serrated Wrack (*Fucus serratus*) began to grow among them and finally take over. Most of these would normally have been eaten down by limpets and top-shells before reaching any size. Now they grew unhindered over almost all the rocky shore, and for the next few years their slippery fronds made walking among them difficult. However, under these seaweeds very large numbers of young limpets appeared, settled out of the plankton in which they had drifted as larvae from spawnings on neighbouring unaffected coasts. This new population of limpets set about eating the abundant seaweeds all around them, and with a plentiful food supply grew fast. They ate the fronds and gnawed at the tough basal stems, thereby so weakening the attachments to the rocks that eventually the weeds became detached and floated away loose, or were torn away by the waves. Little by little the rocks were cleared of the now large and ageing weeds, until by about the mid 1970s the shore had resumed a general appearance of normality.*

All these changes were graphically recorded by shore photography: Alison and I possessed many photographs taken there before the Torrey Canyon struck, not only close-ups of the numerous species living on rocks or in rock-pools, but also views of the shore from many angles. After the stranding of the oil we recorded the

A group of limpets and periwinkles

---

*DPW would now be concerned by the invasive Japweed (*Sargassum muticum*), which is colonising the rock-pools.

hosing of the rocks with detergent and the consequent tragic destruction of the life of a shore we loved. We photographed the flush of green that summer, its suppression by the brown seaweeds in the autumn, their growth in following years, the arrival of the new limpet population, and eventually the clearing away of the brown seaweeds. These photographs, and those taken by others, proved invaluable in assessing what actually happened and the sequence of events as the shore began to recover.

Luckily for the Honeycomb Worm colonies at Duckpool, Torrey Canyon oil never travelled so far eastwards along the north coast of Cornwall. Nevertheless oil spillages are a continuing threat to our marine heritage: we can only hope that the devastation caused in the aftermath of such a tragedy will never occur again.

Alison, Hester and Richard explore a pool at Rhos-on-Sea in 1947. The remains of the old pier, from which DPW trawled for plankton can be seen.

Above: *Honeycomb worm colony* (Sabellaria alveolata) *exposed at low water*
Below: *Underwater view shows heads of worms at the mouths of their sandy tubes*

Above: *Spiny Starfish* (Marthasterias glacialis) *in a Cornish rock-pool*

Snakelocks anemone (Anemonia sulcata) *waves its tentacles*

Photo: Alison Wilson

Above left: *The Furrowed Crab* (Xantho incisus) *may be yellow, green or red-brown*
Right: *Dog whelks* (Nucella lapillus) *and their egg-capsules amid corallines etc, at low tide on a Cornish seashore*

Mussels, limpets and acorn barnacles on mid-tide rocks (Mytilus galloprovincialis, Patella vulgata, Chthamalus stellatus)

Stormy sea at sunset, Tregudda cliffs, North Cornwall, September 1954. Colour photos were not common then, and The Illustrated London News *printed one of the series full page. To obtain it DPW lay full length on the cliffs while Alison clung to the tripod legs.*

CHAPTER SEVEN

Strange Strandings

A few minutes to go and we shall be on the air – live! As we sit around a table facing microphones, our well-rehearsed manuscripts laid out before us, we receive final instructions from our friendly producer, Desmond Hawkins. It is almost 6.30 pm on Thursday, 3 January 1946, and the first broadcast of *The Naturalist* is about to begin. There are four speakers: Geoffrey Grigson the Chairman, Richard Adams to talk about bird-watching, dear old Ludwig Koch to present his recordings of winter bird song, and me. At 6.30 pm precisely the red light comes on and the voice of the curlew, a recording by Ludwig, is heard for the first time in homes throughout the West of England. The notes of the bird die away and the Chairman describes the future scope of this new series of weekly programmes on natural history, each time to begin with the song of the curlew. I am now introduced, and my subject – the Portuguese Man-of-war.

Although I took part in several later programmes of *The Naturalist*, speaking on other subjects, it is that first session which remains most vividly in memory. I cannot help feeling a little personal pride in having had the privilege of presenting the first ever subject in what became a famous long-running series of nature talks. Eventually it all led to the formation in Bristol of the BBC's even more illustrious Natural History Unit and the wonderful television series to be seen on our television screens today.

This was not my first broadcast with the BBC however. Way back in October 1927 I had made a live broadcast to schools from their new Plymouth studio. It was a strange experience. The thickly-carpeted studio had felt-lined walls and heavy purple curtains over the windows; in the middle of the room was the microphone, a box-like arrangement on a heavy stand. I arrived shortly before the appointed hour with my script ticked off into minutes and was greeted by the announcer who stressed the need to keep to time and to pronounce each word distinctly, clipping off its ending properly. It was fortunate that I had my own watch, as the studio clock had stopped (and it was curious that there were several pictures hanging on the wall, but *upside-down* – why I never discovered). A red light came on, I was introduced, the announcer left the room, and I was on the air! A signal-board stood in front of me with a series of little lamps against instructions such as 'Speak louder', 'Speak softer', 'Come nearer', 'Go away', and I kept my eyes on this, my watch, and my script, as I tried to put across my subject, 'Marine Bristle Worms'. But as I read I couldn't help wondering if anyone was listening. I had a horrid vision of a schoolboy less interested in marine worms than in poking the boy in front, and getting hot and bothered I stumbled a little over my words. But no sign lit up. At the end the announcer returned, smoothly signed us off, and assured me that the broadcast had gone through well and clearly.

In 1946 appreciative recognition of *The Naturalist* and my part in it was not long in coming. Next morning in a Bristol hotel I was packing my suitcase when there was a knock on the door and an elderly chamber-maid asked if she could start work on getting the room ready for the next visitor. After exchanging some remarks, probably about the weather, she suddenly looked at me and said 'Didn't I hear you on the wireless last evening?' Assured that she had recognised the voice correctly she told me how very much she and her husband had enjoyed the new programme, so different from anything they had heard before, and how much they were looking forward to the promised programmes to come. Her remarks cheered me tremendously for the train journey back to Plymouth.

Why the Portuguese Man-of-war? Because the previous year it had been much in the news. Through the summer and autumn of 1945, even into early winter, this creature, strange to our waters, had stranded sometimes hundreds at a time on our south-western shores from the Isles of Scilly, along the north coast of Cornwall and on the south and west coasts of Wales. Bathers had been painfully stung by their deep blue tentacles, and so had anyone picking them up, attracted by the brilliant blues and pinks of their bladder floats. Though the Portuguese Man-of-war is a common inhabitant of the ocean surface in tropical and sub-tropical

Portuguese Man-of-war devouring a wrasse

regions throughout the world it is not normally seen off temperate shores. Another obvious question is how does it get its name? Perhaps because in the days of sailing ships making their way south slowly enough for surface creatures to be noticed, the Man-of war, whose scientific name is *Physalia physalis*, was most likely first met with in the latitudes of Portugal. And Man-of-war no doubt because its boat-shaped float is topped by a sail to drive it before the wind trailing its ferocious stinging tentacles which have caused human deaths.

A Portuguese Man-of-war is a strange creature: indeed it is not a single animal at all but a compound colony of individuals, each specialised for one particular function to serve the colony as a whole. First there is the gas-filled float (the gas differs little in composition from air) with a crest set diagonally along the top, functioning as a sail to catch the wind. Attached below the float near the broad end is a cluster of other individuals, sometimes referred to as 'persons'. Among these are the stinging tentacles (each a separate person) to seize and fasten onto fishes coming into contact with them. Then there are the persons specialised for consuming the fishes caught, and others specialised for reproduction. The tentacles are astonishingly extensile, capable of streaming out to distances of many metres, or retracting to only a few centimetres. They are thickly studded with thousands of minute capsules, each containing a coiled hollow thread: on appropriate stimulation, as on contact with a fish (or bather), large numbers of threads are shot out to penetrate the skin and inject a paralysing poison. This stinging apparatus is fundamentally the same as that of jellyfishes and sea-anemones, to which *Physalia* is related, but is more powerful with longer stinging threads. When hundreds of thousands of threads are discharged into a fish, even one as big as a mackerel, it is quickly killed as more and more tentacles latch onto it as it struggles. The tentacles now contract, hauling up their catch to within reach of the feeding persons. These fasten onto the dead fish and the tentacles let go, to stream out once more for further prey. The feeding persons spread wide their mouths, trumpet-shaped fashion, edges touching one another over the fish until it is completely covered by them and digestion begins. Partly-digested particles pass up inside their transparent tubular bodies to nourish the colony as a whole. When much of the fish has been digested the remainder is allowed to fall away.

I was able to watch and photograph this process when in August 1945 a living *Physalia* in first class condition arrived at the Laboratory from the Isles of Scilly, through the kindness of Major A.A.Dorrien-Smith of Tresco. It survived for several days in our tanks, during which time it caught and ate a small wrasse. This fine specimen enabled me to observe and later to publish a more complete account of the behaviour of the species than had previously been known. At that time, unfortunately, I was photographing only in black and white, but nine years later some more Men-of-war sailed up the Channel and several living specimens were brought to the Laboratory. This time I was able to obtain some photographs in colour.

Sailed up is literally true. Although the set of the ocean currents undoubtedly helps, it would seem that these incursions into our waters are due to coincidences of westerly winds and abundance of the species near the northern limits of its range in mid-Atlantic. There are records from 1834 onwards of occasional strandings on our shores of *Physalia* in small numbers, but only in 1862 and 1912 does there appear to have been anything at all comparable with 1945. Then again in October and November of 1954, during long-continued strong westerly winds, very large numbers were cast ashore from the Isles of Scilly eastwards along the Channel coast as far as Kent, with a noteworthy concentratation on the shores of Dorset.

Strange Strandings

There were strandings, too, on the north coasts of Cornwall and Devon, of Wales and even the Isle of Man, though not as abundantly, and we found and photographed a few at Portquin, North Cornwall, in August 1957.

The latter half of 1954 was indeed a memorable year for strandings, memorable also for its wet summer and persistent strong westerly winds. Before the shoals of *Physalia* appeared in the autumn other notable surface dwellers from lower latitudes were driven to our shores. That August, while Alison and I with our children were holidaying at Lower Treneague, we were thrilled one day to find the rocky shore at Trevone sprinkled with fragile purple shells, some empty, others with living snails inside, while some with bubble rafts were floating in the rock-pools. They were Purple Sea-snails (*Ianthina ianthina*), which we had never seen before. We wrote letters to the press appealing for information of these sightings, and the replies we got from other parts of the coast enabled us to map numerous contemporary strandings from the Isles of Scilly, along the north Cornish coast as far as Woolacombe in north Devon. Except for a few on the western side of the Lizard Peninsula there were no reports from south coast districts.

Richard, Hester, and friends search a Trevone rock-pool

Purple Sea-snail photographed in a glass sweet jar

The Purple Sea-snail drifts on the ocean surface suspended below a raft of air-bubbles made with a specialised part of its foot. This lovely snail feeds on another surface drifter, the 'By-the-wind Sailor' (*Velella velella*), a distant relative of the Man-of-war. It would seem that the predator does not pursue its prey, but relies on chance contact as the mixed-up shoals are blown around together. While feeding, the Sea-snail often temporarily abandons its raft, which would get in the way while browsing on *Velella*'s soft underside, relying for the time being on its prey's own skeletal float to keep it from sinking. On the beach that August both predator and prey lay scattered around together.

Without aquarium tanks and the facilities of a laboratory, what were we to do to take advantage of this unexpected offering of rare subjects for the camera? Back in the cottage we found a large glass sweet jar which would do for a tank, and some pie-dishes for views from above. We got sea water from a deep rock-pool and collected a few of the best *Ianthina* and *Velella* and brought them back to the cottage. There we photographed the building of a bubble raft as seen from above, and side views of the Sea-snails floating in the jar, and we made observations on behaviour not previously recorded. One day when we came in we found that the bottom of a basin was covered with gyrating brown specks which a lens showed to be miniature versions of the adults, but still able to swim underwater. While watching the jars twelve-year-old Richard probably became the first person ever to see a Sea-snail give birth, when a little packet shot from its lefthand side, swelling as it sank till suddenly there was a mass of somersaulting young. We needed some kind of pipette but hadn't even a drinking straw so we sent him up the lane to the farm to collect a real one, then sucked up

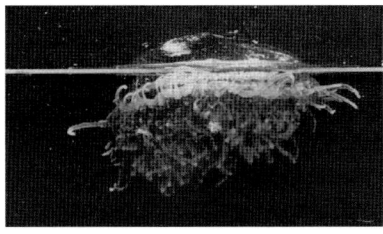

'By-the-wind Sailor'

and preserved some of the babies in methylated spirits – all we had! But none of our *Ianthina* would feed despite being offered some choice living *Velella*.

Velella, unaccompanied by *Ianthina*, is fairly frequently met with floating off our south-western coasts or stranded on shore. Often it is only the dead skeletal remains which are to be picked up from drift-lines, at other times the stranded deep blue living animals arouse the curiosity of holiday makers when they litter sandy beaches. *Velella* has a raft-like float with air-filled buoyancy chambers, made of a sort of horny material. The float is a flattish oval with a stiff upright triangular sail set diagonally across it. In life this skeletal structure is covered with living tissue. Underneath the float a large central mouth opens downwards; it is surrounded by quite a forest of small tubular structures provided with small mouths and bearing on their sides the reproductive organs. Around the edge of the float, also hanging down, is a fringe of tentacles armed with stinging capsules, similar to those of the Portuguese Man-of-war but much feebler, for catching and stunning small prey.

Since that memorable August of 1954 further strandings of *Ianthina* have been reported from time to time, especially in 1964, though we never found them again; we might have done had we been living on the north coast of Cornwall able to visit the shore after every westerly gale. Years of *Velella* strandings occur more often and we have found living specimens stranded on north Cornish shores on several occasions. A truly colossal invasion by this species into our waters took place in June and July 1981 when literally millions were seen from ships as they made their way through mile after mile of them floating at the surface in the entrance to the English Channel. They came ashore in enormous numbers on the west coast of Brittany, on the Isles of Scilly, north and south Cornwall, north Devon and south-west Wales. With them was their predator *Ianthina*, but only in small numbers.

Sea Life in Focus

Ships' Barnacles

Stranded timbers or empty bottles which have been afloat in mid-ocean for a long time are likely to have clusters of Goose or Ships' Barnacles (*Lepas anatifera*) attached below what had been the water-line. These barnacles have long stalks; at their unattached ends are the main bodies protected by limey plates. From between the two sets of plates feathery limbs protrude, sweeping to and fro to enmesh the plankton organisms on which they feed. It was believed in medieval times that stalked barnacles developed into Barnacle Geese, birds which breed in the then largely unknown Arctic, for which reason their nests and eggs had never been seen. The mysterious appearance of these geese on European shores in winter seemed therefore to be explained by a metamorphosis of the stalked barnacles into geese. The shape and colouring of the protecting body plates suggested to the medieval mind a Barnacle Goose's wings. The plates were later supposed to become feathered. When printing had been invented, drawings in books illustrated the remarkable process. Some show a branched barnacle tree overhanging water and shedding birds into it. The other common name derives from this barnacle often attaching to the hulls of ships, seriously impeding their

Medieval 'Barnacle Goose Tree'

progress through the water in the days of sail.

Yet another kind of oceanic stalked barnacle is sometimes stranded on western shores at the same time as *Physalia* and *Velella*. This is the Buoy-making Barnacle (*Lepas fascicularis*) which when young attaches to floating scraps of seaweed and other debris, secreting a frothy bubble float as it grows bigger and heavier. Several individuals settle together to form a cluster, and like the other species they feed on plankton. Both kinds of stalked barnacles produce free-swimming larvae which after development in the plankton eventually attach to objects floating far distant from those of their parents, so ensuring dispersal of the species.

Turtles, usually young ones, are sometimes brought to our shores by persistently strong winds. In August 1945, when the Portuguese Man-of-war was stranding along the north coast of Cornwall, three small Loggerhead Turtles (*Caretta caretta*) were found with them at Hayle and at Bude. Six others were subsequently found on shores of Ireland, Wales, the Hebrides and even as far north as the Shetlands. Quite an invasion of this tropical marine reptile within only a few months. The young turtles must have wandered in the Atlantic unusually far to the north at a time when strong winds were blowing continuously towards the British Isles, hurrying them along in our direction. Records of turtles reaching our shores refer usually to single individuals and are spaced years apart, but it is not too surprising to find them.

Young Ridley's or Kemp's Turtle

Sea Life in Focus

Nowadays I myself am stranded high and dry well above the high water mark. Rarely these last few years have I been able to venture onto the shores, and then not alone in case I stumble. As, well inland, I write from vivid memories, old notes, old letters, old apparatus, and many, many photographs, my mind traverses the past from early days to the present time. I recall the advances in photographic technology as the years went by, the steady improvements in sensitive materials, in cameras and in flash-light apparatus, making the work of the photographer ever easier and more effective, making better pictures possible, until now it can almost be said that modern apparatus and trade processing do most of the work. These days anybody who can aim and hold a camera steady can take technically good photographs in colour without any knowledge whatever of the fundamentals of the photographic process, and with quite cheap cameras too. I do not mean to imply that all nature photography is now easy, but what I do feel is that the modern photographer is relieved of many chores, inevitable in the old days, and has been freed to devote his time, energies and patience to finding and approaching his subjects, whether on land or under the sea. The actual taking of the photographs, apart from fixing the camera to point in the right direction and focussing (and sometimes not even that) can be left to electronic mechanisms which calculate the exposure and synchronise any flash which may be required. Of course it can still be hard work, patience and time consuming, especially in the wilder regions of the world, which can be dangerous too. Nevertheless, photography of seashore life, of animals in aquarium tanks, even of living plankton organisms under the microscope, previously so difficult and years ago so rarely undertaken, has become almost too easy; many do it and do it well. I sometimes wonder whether, if I were again a young marine biologist at the beginning of my career, I should nowadays be tempted to devote much time to taking photographs. Perhaps I should only do so to record my researches. Without the technical difficulties of the old days the challenge to invent and construct the apparatus to overcome them is no longer there; the excitement of personal achievement and much of the fun have gone. Or have they? Perhaps I am too old to get excited that way any more and Alison is, alas, no longer with me to stimulate and help. But it still gives pleasure to relive the challenges in my thoughts, as I have done while writing these pages, recording something of my photographic methods and experiences during more than sixty years.

Sun-star

Afterword

In his last years DPW was greatly handicapped by arthritis, but with the same ingenuity that he had brought to his early photography he designed himself a wheeled 'gadabout' on which he was able to whizz round his kitchen and continue to live independently.

The disappointment of not finding a publisher for his memoirs in 1987 was mitigated by the thrill of hearing in the same year that he had been awarded the Honorary Fellowship of the Royal Photographic Society. The citation reads:

> For his pioneering work begun in the 1920s on the development of flash photography of living marine life including planktonic organisms. He was the first person successfully to photograph live plankton with a microscope. For nearly half a century his photographs continued to be featured in popular books on marine life, as well as appearing in his own scientific papers.

Unfortunately he suffered a minor stroke before the ceremony in Bath, but the RPS President, Heather Angel, came to the house to present the certificate. Afterwards, framed, it became his proudest possession.

Following his sudden death in December 1991 there were a number of tributes. An obituary in the *Guardian* was headed 'Through a lens deeply' and illustrated with the sea horses photo used in this book as a frontispiece. The *Times* obituary began 'Douglas Patrick Wilson became internationally known both for his studies on the factors controlling the development and metamorphosis of marine invertebrate larvae and for his beautiful photographs of marine life', while Heather Angel wrote in the *Journal of the Royal Photographic Society* that 'Douglas Wilson's photographs were not only accurate in every scientific respect, they were also beautiful studies of subjects rarely seen by the average person'. Such comments would have given him tremendous pleasure.

My father dedicated his memoirs to his grandchildren and to the memory of his wife. I should like to dedicate my efforts in seeing them finally into print to his memory.

HD

Photo: Catherine Forrest

Douglas and Alison Wilson at Double Waters on Dartmoor, September 1973